Oceans

Environmental Issues, Global Perspectives

Oceans

James Fargo Balliett

Routledge
Taylor & Francis Group

LONDON AND NEW YORK

First published 2010 by M.E. Sharpe

Published 2015 by Routledge
2 Park Square, Milton Park, Abingdon, Oxon OX14 4RN
711 Third Avenue, New York, NY 10017, USA

Routledge is an imprint of the Taylor & Francis Group, an informa business

Library of Congress Cataloging-in-Publication Data

Balliett, James Fargo.
Oceans: environmental issues, global perspectives / James Fargo Balliett.
     p. cm.
Includes bibliographical references and index.
ISBN 978-0-7656-8229-1 (hardcover: alk. paper)
1. Ocean—Environmental aspects. 2. Global environmental change. I. Title.

GC21.B34 2010
333.91'64—dc22   2010019428

Figures on pages 7, 9, 13, 84, and 98 by FoxBytes.

ISBN 13: 9780765682291 (hbk)

Contents

Over the past 150 years, the planet Earth has undergone considerable environmental change, mainly as a result of the increasing number of people living on it. Unprecedented population growth has led to extensive development and natural resource consumption. A population that numbered 978 million people worldwide in 1800 reached 6.7 billion in 2009 and, according to the United Nations, may well exceed 8 billion by 2028.

This sixfold growth in population has brought about both positive and negative outcomes. Developments in medicine, various natural and social sciences, and advanced technology have resulted in widespread societal improvements. One measure of this success, average life expectancy, has climbed substantially. In the United States, life expectancy was 39.4 years in 1850; by 2009, it had grown to 77.9 years.

Another major change for global nations and cultures has been the accessibility and sharing of information. Though once isolated by oceans and geography, few communities remain untouched by technological innovation. The construction of jumbo jet planes, the development of advanced satellites and computers, and the power of the Internet have made travel and information readily available to an unprecedented number of people.

Technological advances have allowed residents of just about anywhere on the planet to share events and information almost instantly. For example, images of an expedition on the summit of Mount Everest or a scientific investigation in the middle of the Atlantic Ocean can be posted online via satellite and viewed by billions of citizens worldwide. In addition, cumulative and individual environmental impacts now can be assessed faster and more comprehensively than in years past.

During the same period, however, regulated and unregulated residential and business development that consumes natural resources has had profound impacts on the global environment. Such impacts are evident in the clear-cutting or burning of forests, whether this deforestation is for fuel or building materials, or simply to clear the land for other activities; the drainage of wetlands to divert freshwater, expand agriculture, or provide more building upland; the overfishing of oceans to meet an ever-growing appetite for seafood; the stripping of mountaintops for fuel, metals, or minerals; and the pollution of freshwater sources by the waste output of industrial and residential communities. In fact, pollution has spread across land, air, and water biomes in ever-increasing concentrations, causing considerable damage, especially to fragile ecosystems.

The Emergence of a Global Perspective

Environmental Issues, Global Perspectives provides a fresh look at critical environmental issues from an international viewpoint. The series consists of five individual volumes: *Wetlands, Forests, Mountains, Oceans,* and *Freshwater.* Relying on the latest accepted principles of science—the acquisition of knowledge based on reasoning, experience, and experiments—each volume presents information and analysis in a clear, objective manner. The overarching goal of the series is to explore how human population growth and behavior have changed the world's natural areas, especially in negative ways, and how modern society has responded to the challenges these changes present—often through increased educational efforts, better conservation, and management of the environment.

Each book is divided into three parts. The first part provides background information on the biome being discussed: how such ecosystems are formed, the relative size and locations of such areas, key animal and plant species that tend to live in such environments, and how the health of each biome affects our planet's environment as a whole. All five titles present the most recent scientific data on the topic and also examine how humans have relied on each biome for survival and stability, including food, water, fuel, and economic growth.

The second part of each book contains in-depth chapters examining seven different geographically diverse locations. An overview of each area details its unique features, including geology, weather conditions, and endemic species. The text also examines the health of the natural environment and discusses the local human population. Short- and long-term environmental impacts are assessed, and regional and international efforts to address interrelated social, economic, and environmental issues are presented in detail.

The third part of each book studies how the cumulative levels of pollution and aggressive resource consumption affect each biome on a global scale. It provides readers with examples of local and regional impacts—filled-in wetlands, decimated forests, overdeveloped mountainsides, empty fishing grounds, and polluted freshwater—as well as responses to these problems. Although each book's conclusion is different, scenarios are highlighted that present collective efforts to address environmental issues. Sometimes, these unique efforts have resulted in a balance between resource conservation and consumption.

The *Environmental Issues, Global Perspectives* series reveals that—despite the distance of geography in each title's case studies—a common set of human-induced ecological pressures and challenges turns up repeatedly. In some areas, evidence of improved resource management or reduced environmental impacts is positive, with local or national cooperation and the application of new technology providing measurable results. In other areas, however, weak laws or unenforced regulations have allowed environmental damage to continue unchecked: Brazen frontiersmen continue to log remote rain forests, massive fishing trawlers still use mile-long nets to fill their floating freezers in the open ocean, and communities, businesses, vehicles, factories, and power plants continue to pollute the air, land, and water resources. Such existing problems and emerging issues, such as global climate change, threaten not just specific animal and plant communities, but also the health and well-being of the very world we live in.

Wetlands

Wetlands encompass a diversity of habitats that rely on the presence of water to survive. Over the last two centuries, these hard-to-reach areas have been viewed with disdain or eliminated by a public that saw them only as dangerous and worth-less lowlands. The *Wetlands* title in this series tracks changing perceptions of one of the world's richest and biologically productive biomes and efforts that have been undertaken to protect many areas. With upland and coastal development resulting in the loss of more than half of the world's wetlands, significant efforts now are under way to protect the 5 million square miles (13 million square kilometers) of wetlands that remain.

In *Wetlands*, three noteworthy examples demonstrate the resilience of wetland plants and animals and their ability to rebound from human-induced pressures: In Central Asia, the Aral Sea and its adjacent wetlands show promising regrowth, in part because of massive hydrology projects being implemented to undo years of damage to the area. The Everglades wetlands complex, spanning the lower third of Florida, is slowly reviving as restorative conservation measures are implemented. And Lake Poyang in southeastern China has experienced increased ecological health as a result of better resource management and community education programs focused on the vital role that wetlands play in a healthy environment.

Forests

Forests are considered the lungs of the planet, as they consume and store carbon dioxide and produce oxygen. These biomes, defined as ecological communities dominated by long-lived woody vegetation, historically have provided an economic foundation for growing nations, supplying food for both local and distant markets, wood for buildings, firewood for fuel, and land for expanding cities and farms. For centuries, industrial nations such as Great Britain, Italy, and the United States have relied on large tracts of forestland for economic prosperity.

The research presented in the *Forests* title of this series reveals that population pressures are causing considerable environmental distress in even the most remote forest areas. Case studies provide an assessment of illegal logging deep in South America's Amazon Rain Forest, a region closely tied to food and product demands thousands of miles away; an examination of the effect of increased hunting in Central Africa's Congo forest, which threatens wildlife, especially mammal species with slower reproductive cycles; and a profile of encroachment on old-growth tropical forests on the Southern Pacific island of Borneo, which today is better managed, thanks to the collective planning and conservation efforts of the governments of Brunei, Indonesia, and Malaysia.

Mountains

Always awe-inspiring, mountainous areas contain hundreds of millions of years of history, stretching back to the earliest continental landforms. Mountains are characterized by their distinctive geological, ecological, and biological conditions. Often, they are so large that they create their own weather patterns. They also store nearly one-third of the world's freshwater—in the form of ice and snow—on their slopes. Despite their daunting size and often formidable climates, mountains are affected by growing local populations, as well as by distant influences, such as air pollution and global climate change.

The case studies in the *Mountains* book consider how global warming in East Africa is harming Mount Kenya's regional population, which relies on mountain runoff to irrigate farms for subsistence crops; examine the fragile ecology of the South Island mountains in New Zealand's Southern Alps and consider how development threatens the region's endemic plant and animal species; and discuss the impact of mountain use over time in New Hampshire's White Mountains, where stricter management efforts have been used to limit the growing footprint of millions of annual visitors and alpine trekkers.

Oceans

Covering 71 percent of the planet, these saline bodies of water likely provided the unique conditions necessary for the building blocks of life to form billions of years ago. Today, our oceans continue to support life in important ways: by providing the largest global source of protein in the form of fish populations, by creating and influencing weather systems, and by absorbing waste streams, such as airborne carbon.

Oceans have an almost magnetic draw—almost half of the world's population lives within a few hours of an ocean. Although oceans are vast in size, exceeding 328 million cubic miles (1.37 billion cubic kilometers), they have been influenced by and have influenced humans in numerous ways.

The case studies in the *Oceans* title of this series focus on the most remote locations along the Mid-Atlantic Ridge, where new ocean floor is being formed 20,000 feet (6,100 meters) underwater; the Maldives, a string of islands in the Indian Ocean, where increasing sea levels may force residents to abandon some communities by 2020; and the North Sea at the edge of the Arctic Ocean, where fishing stocks have been dangerously depleted as a result of multiple nations' unrelenting removal of the smallest and largest species.

Freshwater

Freshwater is our planet's most precious resource, and it also is the least conserved. Freshwater makes up only 3 percent of the total water on the planet, and yet the majority (1.9 percent) is held in a frozen state in glaciers, icebergs, and polar ice fields. This leaves only 1.1 percent of the total volume of water on the planet as freshwater available in liquid form.

The final book in this series, *Freshwater,* tracks the complex history of the steady growth of humankind's water consumption, which today reaches some 3.57 quadrillion gallons (13.5 quadrillion liters) per year. Along with a larger population has come the need for more drinking water, larger farms requiring greater volumes of water for extensive irrigation, and more freshwater to support business and industry. At the same time, such developments have led to lowered water supplies and increased water pollution.

The case studies in *Freshwater* look at massive water systems such as that of New York City and the efforts required to transport this freshwater and protect these resources; examine how growth has affected freshwater quality in the ecologically unique and geographically isolated Lake Baikal region of eastern Russia; and study the success story of the privatized freshwater system in Chile and consider how that country's water sources are threatened by climate change.

Acknowledgments

I owe the greatest debt to my wonderful Mom and Dad, Nancy and Whitney, who led me to the natural world as a child. Thank you for encouraging curiosity and creativity, and for teaching me to be strong in the midst of a storm. I also could not have gotten this far without steady support, expert advice, and humorous optimism from my siblings: Blue, Julie, Will, and Whit.

I greatly appreciate the input of Dr. Arri Eisen, Director of the Program in Science and Society at Emory University, at key stages of this project. My sincere thanks also go to the superb team at M.E. Sharpe, including Donna Sanzone, Cathy Prisco, and Laura Brengelman, as well as Gina Misiroglu, Jennifer Acker, Deborah Ring, Patrice Sheridan, and Leslee Anderson. Any title that explores science and the environment faces daunting hurdles of ever-changing data and a need for the highest accuracy. This series benefited greatly from their precise work and steady guidance.

Finally, *Environmental Issues, Global Perspectives* would not have been possible without the efforts of the many scientists, researchers, policy experts, regulators, conservationists, and writers with a vested interest in the environment.

The last few decades of the twentieth century brought a significant change in awareness and attitudes toward the health of this planet. Scholars and laypeople alike shifted their view of the environment from something simply to be consumed and conquered now to a viewpoint of it as a significant asset because of its capacities for such measurable benefits as flood control, water filtration, oxygen creation, pollution storage and processing, and biodiversity support, as well as other positive features.

Knowledge of Earth's finiteness and vulnerability has resulted in substantially better stewardship. My thanks go to those people who, through their visions and hard work, have taught the next generation that fundamental science is essential and that humankind's collective health is inextricably tied to the global environment.

James Fargo Balliett
Cape Cod, Massachusetts

INTRODUCTION
TO OCEANS

1   Oceans Around Us

Oceans are all around us. Of the 6.8 billion people who live on planet Earth, some 44 percent lives within 90 miles (145 kilometers) of an ocean. These vast bodies of salt water cover 71 percent of the planet and hold an estimated volume of 328 million cubic miles (1.37 billion cubic kilometers), or 343,423,668,428,484,681,262, or over 343 quintillion, gallons (approximately 1.4 sextillion liters, or the number 1.4 followed by 21 zeroes).

Oceans are a distinct part of human evolution and history. They were a source for some of the first forms of life and today support it in multiple ways, primarily through the maintenance of temperature equilibrium across the globe, the formation and influencing of weather patterns, and the long-term storage of waste streams, including carbon dioxide. The ocean environment also hosts a number of threats to humans, such as tsunamis, earthquakes, and storms.

Ocean Formation

The events leading to the creation of the world's oceans date back to when the Earth was in its early stage of formation. Roughly 4.3 billion years ago, the molten surface of the planet began to cool into a landscape of sunken lowlands, mountains, and uplands. Vast plumes of hydrogen and methane steam that had been trapped underground were driven by heat and pressure into the atmosphere. This steam was laden with minerals such as calcium, chlorine, magnesium, and sodium sulfate, as well as other chemicals. The Earth also was impacted by thousands of large and small meteors that deposited considerable amounts of rock, heavy with water vapor, onto the planetary surface.

As clouds of gaseous materials collected above the land, the first atmospheric conditions triggered rainstorms. These lasted, off and on, for millions of years, and the first large bodies of water formed.

Rivers on higher ground drained into the lower lands, eventually filling into ocean-sized bodies. Massive amounts of sediment and minerals were washed downward as well. This downward flow and that of rainwater, which mixed with dissolved salts from bedrock, are why the oceans are saline.

Deep in submerged canyons were underwater volcanic vents releasing lava and plumes of carbon dioxide and hydrogen. It was here, in an underwater environment, about 3.5 billion years ago, that the building blocks of living creatures most likely formed.

This ocean event is key to abiogenesis, the study of how life on Earth arose out of inanimate chemicals. Prevalent volcanic discharge, rich in ammonia and

HYDROTHERMAL ORIGIN OF LIFE

German chemist Günther Wächtershäuser proposed in 1997 that the following conditions allowed for the creation of the building blocks of life:

1. Water environment, preferably saline

2. Temperatures between 100 and 250 degrees Fahrenheit (38 and 121 degrees Celsius)

3. Presence of iron, nickel, and cobalt sulfide

4. High pressure of a deepwater environment

5. Presence of superheated carbon monoxide

Wächtershäuser concluded that in these conditions, certain metals catalyze the binding of basic carbon molecules into more complex molecules. Mineral-based chemistry, in the right conditions, forms biologically active organic life. While this theory shows promise, the exact source of life on Earth may never be known, because these events occurred billions of years ago and cannot be proven.

hydrogen sulfide, mixed with salt water and altered floating metal particles such as iron and nickel. These organic compounds formed small molecules, which eventually combined to form chains of amino acids. These amino acids played a role in the eventual creation of single-celled animals, many of which evolved into oceanic multicellular organisms, including plankton, plants, fish, and other animals.

By 500 million years ago, the oceans had reached a size similar to that of today. Even though water now covered 71 percent of the planet, the land under the water still was in a state of constant change, being altered by a range of geologic forces. Fifteen major tectonic plates (and up to forty-one smaller ones) served as the outer rock layer on the planet.

In 1912, the German scientist Alfred Wegener hypothesized that, around this time, the continental plates had formed into one mass, called Pangaea (meaning "ancient Earth" in Greek). This single amalgam of land straddled the equatorial region of the Earth, surrounded by a superocean that Wegner called Panthalassa (meaning "all seas" in Greek).

About 20 million years later, Wegener theorized, Pangaea rifted, or split, and new ocean floor began forming. Pangaea's pieces—the Earth's continents—have been moving away from each other ever since. It took half of a century for another scientist, American oceanographer William Maurice Ewing, to prove Wegener's hypothesis of continental drift.

Beginning in the 1950s, Ewing extensively surveyed the ocean bottom. His use of sonar waves to "see" the contours of the ocean ridges and slopes revealed an

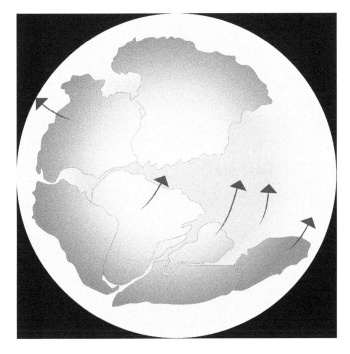

This diagram illustrates the Earth's continental drift, including the fragmented Pangaea, or "ancient Earth" that existed 500 million years ago. Since then, the plates broke apart and have moved to their current formation at a rate of up to 6 inches (15.2 centimeters) per year. *(DEA/ D'Arco Editori/De Agostini/Getty Images)*

active massive undersea ocean rift—a large underwater mountain range emerging from a divergent boundary between two tectonic plates—at the bottom of the Atlantic Ocean. Ewing recorded multiple earthquakes and the presence of new magma emerging from the rift itself, forming undersea mountain ranges. Ewing also recorded the collisions of plates of varying densities in other locations, where they formed massive subduction zones; bedrock was compacted and then pressed back into the planetary crust, where it eventually was remelted into magma.

This cycle of formation and destruction of continents and ocean floor has occurred multiple times over the last 2 billion years. In another 30 million years, the Atlantic Ocean will be larger, the Pacific Ocean will be smaller, and a number of other oceanic tectonic changes will have occurred.

The Physical and Chemical Oceans

Although there are five distinct oceans (Arctic, Atlantic, Indian, Pacific, and Southern), these bodies of salt water are not separate. They combine to form what scientists term the World Ocean, a global expanse that covers about 139 million square miles (360 million square kilometers). The World Ocean is approximately 300 times larger than all upland combined. Given this proportion, this planet might have more aptly been named "Oceanus" and not "Earth." In fact, if the entire surface of the planet were flat, ocean water would cover the surface to a depth of 12,000 feet (3,658 meters).

More than 60 percent of the Earth's total landmass is found in the Northern Hemisphere, while less than 40 percent exists in the Southern Hemisphere. More than half of the oceans exceed 10,000 feet (3,048 meters) in depth, and the average depth is 12,430 feet (3,789 meters). The deepest known location, in the Mariana Trench in the western Pacific Ocean, is approximately 36,201 feet (11,034 meters) underwater.

All oceans are composed of salt water, which makes up 97 percent of the Earth's water supply. This water has a salinity content that ranges between 3.1 and

GLOBAL OCEAN SIZES

Ocean	Size (Million Square Miles)
Arctic	5.44
Atlantic	33.50
Indian	26.90
Pacific	65.60
Southern	7.85
Total:	139.29

Ocean Zones

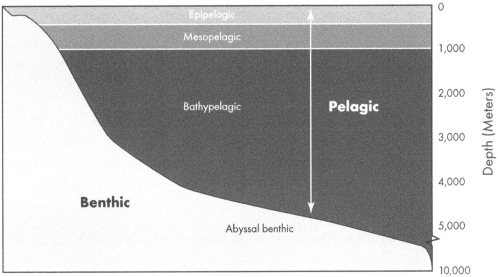

3.8 percent and is rarely uniform, especially where freshwater and salt water mix at river mouths, at estuaries (semi-enclosed coastal water bodies), or near glaciers or ice floes. The pH (measure of acidity) of ocean water ranges from 7.8 to 8.5.

Seawater is composed of a number of elements that are dissolved in varying concentrations. A common formulation contains 85.4 percent oxygen, 11.5 percent hydrogen, 1.9 percent chlorine, 1 percent sodium, 0.09 percent sulfur, 0.04 percent calcium, 0.04 percent potassium, 0.006 percent bromine, and 0.002 percent carbon.

Humans cannot survive on salt water alone, because the ingested concentration of sodium chloride requires additional freshwater to process and eliminate it, resulting in dehydration. Ocean fish, birds, mammals, and other sea creatures have evolved to manage the salty conditions, and their bodies can eliminate salts.

Oceans host unique underwater environments that vary, depending on their depth, bottom surface conditions and composition, water temperature, currents, and proximity to land. The ocean, as viewed in a cross section, is divided into multiple horizontal and vertical zones. The two main ones are divided horizontally and include the benthic realm, which forms the ocean floor environment, and the pelagic realm, which includes any area above the ocean floor.

Life in Oceans

Approximately 3.5 billion years ago, salty warm waters hosted a unique habitat, as they were surrounded by erupting volcanoes and skies filled with an atmosphere made from a combination of sulfur, carbon dioxide, nitrogen, and methane (which would be highly poisonous to any oxygen-dependent animals, such as humans). Despite these conditions, some of the microbial species, known as phytoplankton, evolved over millions of years to eventually take in carbon dioxide and produce oxygen as a by-product. This critical contribution helped create an environment where a broader diversity of oxygen-dependent water and land species could exist.

Diatoms were one of the first types of phytoplankton to emerge. These microscopic plants float at or near the surface of the water, relying upon sunlight and photosynthesis to grow and reproduce. Diatoms are the foundation of the marine food chain; they are a food source for zooplankton (larger animal species) and small fish.

Today, there are thirty-four major types, or phyla, of different organisms that live in the ocean, including Angiospermophyta (seagrasses), Arthropoda (crabs, shrimp, and barnacles), Clorophyta (green algae), Chordata (fish and aquatic mammals), Coelenterata (corals and jellyfish), Echinodermata (sea urchins and star fish), Mollusca (clams, octopuses, and oysters), and Porifera (sponges). Ocean life is called a food web because each species consumes others to survive.

The ocean food web begins with the smallest creatures such as algae and plants; they are primary producers that multiply to provide a massive source of food for larger species. As the species increase in size, their numbers and density also shrink, and fewer species eat them.

The largest ocean creature, the blue whale (*Balaenoptera musculus*), which grows to a length of 110 feet (34 meters) and a weight of 200 tons, relies on krill (small shrimp-like invertebrates in the zooplankton family) to survive. Given their considerable size, these whales may eat up to 8,000 pounds (3,629 kilograms) of krill in a single day. They filter food into their mouths with baleen plates. This material is fine enough to capture tiny krill while releasing water.

One of the ocean's most significant phyla is the Chordata, which consists of a broad subgroup (also called a "class") of noteworthy animals, including bony fishes (class Osteichthyes), fish with cartilage (class Chondrichthyes), birds (class Aves), mammals (class Mammalia), tunicates, or "sea squirts" (class Urochordata), and reptiles (class Reptilia). About 570 million years ago, an unprecedented evolution occurred in marine organisms in the Chordata family; they developed the first versions of a spinal cord.

This feature helped form a stronger body, as well as a central nervous system capable of higher intelligence. Species with more sensory perception and ability were better suited to survive and thrive in diverse environments. Chordata's evolution has led to considerable biological improvement and cognitive development over time.

Ocean Geology

Approximately 5 billion years ago, the Earth was a large cloud of rocks and gas barely forming a rounded shape. Four hundred million years later, the mass had mostly solidified into a planet and developed a defined orbit around the sun.

Intense pressure and heat had formed the globe into four distinct layers: an inner core composed of solid and semisolid nickel iron that is 750 miles (1,207 kilometers) in diameter; an outer core made of liquid nickel iron that is 2,200 miles (3,540 kilometers) in diameter; a mantle that consists of iron, magnesium, and silicon and stretches for 1,800 miles (2,896 kilometers); and a crust, which is only 2 to 5 miles (3 to 8 kilometers) in thickness. It is this crust, consisting of multiple tectonic plates, that forms today's ocean floor.

While geologists have determined that the oldest continental plates are 4.2 billion years old, ocean plates are not older than 200 million years. This fact is due to the ocean floor's constant cycle of destruction and creation. Several mid-ocean ridges create new floor, and subduction zones crush the existing floor, with the result that the age of the ocean floor's rocks is relatively young.

Ocean floor is formed, altered, or destroyed in four ways: volcanic events; the collision of continental and ocean plates; the collision of two ocean plates; and

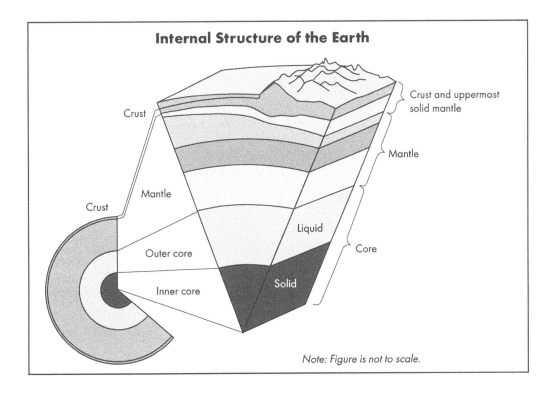

Internal Structure of the Earth

Crust and uppermost solid mantle

Crust

Mantle

Crust

Mantle

Outer core

Liquid

Core

Inner core

Solid

Note: Figure is not to scale.

the activity of mid-ocean ridges. The movement of ocean tectonic plates happens at a very slow rate of only a few inches per year.

Volcanic Events

Volcanoes are common on the ocean floor. Occasional fractures in the crust allow a vein of magma to push up from under a tectonic plate and collect on the surface. Even with an ocean's water weight resting on the floor, the lava emerges to form a structure.

One type of structure is called a hot spot volcano, which sometimes breaks the surface of the water and forms a series of islands. When the tectonic plate moves, the magma vein reappears, forming another volcano, or island, a short distance away. This type of volcano created the Aleutian Islands, a chain of more than 300 small volcanic islands in the northern Pacific Ocean, approximately 1,200 miles (1,931 kilometers) west of the Alaska Peninsula, and the Galapagos Islands, an archipelago of volcanic islands distributed around the equator in the Pacific Ocean, approximately 604 miles (972 kilometers) west of Ecuador. Scientists have identified approximately fifty current hot spots worldwide.

Continental and Ocean Plates

The collision of an ocean and a continental plate is another method by which the ocean floor is altered. The convergence of these two types of plates typically ends

Observations of wildlife of the Galapagos Islands played a central role in the theories of Charles Darwin (1809–1882) regarding evolution and natural selection. Darwin's research during a visit to the islands in 1835 was instrumental in the writing of his landmark book *The Origin of Species*. (© Alexander/Fotolia)

up with the thinner, denser ocean plate being pushed downward, or subducted, below the thicker and lighter continental plate. The ocean plate is crushed into a deep trench, and the material it is made of eventually is forced under the crust where it is melted down to form new lava.

Another scenario occurs when ocean plates that consist of denser rock come into contact with sections of continental rock, especially if the ocean plates are larger or have more momentum than the land ones. The ocean plates can apply enough pressure to landmasses to force an upward lift. The result is mountain ranges such as the Andes Mountains in South America and the Himalayan Mountains in India and other parts of Asia.

Two Ocean Plates

When one ocean plate runs into another, the outcome is similar to that of a land and ocean plate collision. The lighter plate typically is lifted up, while the denser plate is compacted and crushed.

The Mariana Trench, a 1,580-mile-long (2,542-kilometer-long) trench located to the east of the Mariana Islands, south of Japan and north of New Guinea, was formed when the Pacific Plate collided with the Philippine Plate. The deepest part of the trench is 6.8 miles (10.9 kilometers) deep. Sonar surveys of the ocean bottom show the trench as an inverted C curve, with layers of the Philippine Plate climbing up over the denser Pacific Plate.

Mid-Ocean Ridges

A mid-ocean ridge is a lengthy underwater mountain range that has been created by magma emerging between two tectonic plates. While volcanoes may emerge in an individual location, a mid-ocean ridge may measure thousands of miles in length. Up to a dozen of these ridges exist around the world.

Most mid-ocean ridges are part of a 40,000-mile-long (64,360-kilometer-long) formation that runs through portions of the Arctic, Atlantic, Indian, Pacific, and Southern oceans. Smaller ridges can be found where the Pacific Ocean meets the South American coast (Galapagos Rise) and where the southern Pacific Ocean meets the Antarctic Plate (Pacific-Antarctic Ridge).

Movement of Ocean Water

Water within the oceans moves in multiple directions at varying levels, from steady surface currents heated directly by the sun to powerful deep-ocean currents cooled by the flow of water from arctic regions. All ocean life relies on water movement—which delivers essential food and oxygen—to survive. Moving water also influences weather patterns and, combined with winds, forms storms that play a central role in how and where salt water travels across ocean basins.

Currents

Two main types of currents exist in the open ocean: surface and deepwater. The surface currents, also called gyres, flow down to 600 feet (183 meters) deep in similar patterns around the world.

In the north, a clockwise flow of surface water exists in the Atlantic and Pacific oceans. In the Southern Hemisphere, this pattern reverses to a counterclockwise fashion in the Atlantic, Pacific, and Indian oceans. These circular gyres, the motion of which is explained through a theory known as the Coriolis effect, are caused by the gravitational rotation of the Earth and result in objects (such as wind currents and moving water) being deflected (to the right in the north and to the left in the south) due to the inertia of planetary motion.

A singular deep ocean current, reaching down to 13,000 feet (3,962 meters), stretches for up to 42,000 miles (67,578 kilometers) across the globe. Named the Great Ocean Conveyor, this slow-moving flow relies on heated waters to continually move oxygenated water from the surface down to the bottom of the oceans, where these waters then uplift nutrients from the depths to the surface. The Great Ocean Conveyor loops its way around the planet, moving up to 1 billion cubic feet (30 million cubic meters) of water every second, yet taking up to 1,000 years to complete a circuit.

This process begins in the southern Pacific Ocean, where warm shallow water is transported through the Indian Ocean and up along the west coast of Africa. When it reaches the cold Arctic Ocean, water from the depths is carried south

OCEAN CURRENTS
(listed from top to bottom)

Name	Depth
Surface water	Down to 600 feet
Central water	From surface water to where temperature changes
Intermediate water	Down to 5,000 feet
Deep water	Down to 13,000 feet
Bottom water	In contact with the ocean floor (up to 36,000 feet)

Source: Pernetta, John. Guide to the Oceans. Buffalo, NY: Firefly, 2004.

Saltwater Currents: Great Ocean Conveyor Belt

along the eastern side of South America. It eventually reaches the Antarctic coast and finally flows back into the southern Pacific Ocean, where the process starts all over again.

Both surface and deep-ocean currents move around the planet through areas that can be divided into temperature or climate zones. There are four key zones:

1. Two bands of polar water (less than 40 degrees Fahrenheit, or 4.4 degrees Celsius) exist at the top and bottom of the planet.

2. There are two bands of sub-polar cold water (40 to 50 degrees Fahrenheit, or 4.4 to 10 degrees Celsius).

3. There are two bands of warmer temperate water (50 to 69 degrees Fahrenheit, or 10 to 21 degrees Celsius).

4. One band of tropical water (above 69 degrees Fahrenheit, or 21 degrees Celsius) runs along the equator.

One example of local current is an upwelling. These currents, often found in coastal areas, bring nutrients from the ocean bottom to the surface. This phenomenon occurs where a continent or island sits next to cold, deep waters that are rich in phosphate, nitrate, dissolved carbon, and other organic materials.

The nutrients brought to the surface are a boon for microscopic life, such as blooms of plankton, which are fed by nutrients in the water and sunlight. Small fish such as herring (*Clupea sp.*) and anchovies (*Engraulidae sp.*), thrive on the abundance of plankton. In turn, the small fish are eaten by bigger fish, such as tuna (*Thunnus sp.*) and salmon (*Oncorhynchus sp.*).

Common upwellings around the world include the waters off the states of California, Oregon, and Washington in North America, southern Brazil in South America, and eastern New Zealand in the southern Pacific Ocean.

Waves

Waves are a part of surface ocean currents. Visible in any body of water, a wave is an example of energy traveling from one point to another. Although the sun's rays and an earthquake's displacement of tectonic rock can create waves, wind is the most common wave maker.

There are two common types of waves, which vary in intensity and duration. Small, or capillary, waves last less than a second and have a length of only a few centimeters. Sea waves range in size from a few inches to a few hundred feet, last for minutes, hours, or days, and may travel thousands of miles.

Although a wave appears as a crest or ridge of water moving across the surface, it actually is a circular motion—again, most often created by the wind—that extends under the surface. Only when the wave comes into contact with the ocean bottom in shallow water is it forced up to break on the surface. Some areas of the planet experience higher waves than others; this generally happens when shallow water is adjacent to much deeper water.

While storm wave heights commonly reach 25 feet (8 meters), some grow to exceed 50 feet (15 meters) in rare conditions. A few examples of areas that boast these types of waves include the southern Indian Ocean, major parts of the Southern Ocean near Antarctica, the southern tip of South America where the Atlantic and Pacific oceans meet, and the central Pacific Ocean off the island of Hawaii.

Giant sea waves, also called rogue waves, have been measured at 100 feet (30 meters) above the surface. In 1995, a laser on an oil-drilling platform in Europe's North Sea recorded a rogue wave 85 feet (26 meters) high. In 2001, a satellite survey by the European Space Agency (ESA), Europe's multi-country space research and policy organization, recorded three weeks of wave activity in the southern Atlantic Ocean, during which it measured ten 82-foot (25-meter) waves. According to the ESA, approximately 200 supertankers and container ships exceeding 656 feet (200 meters) in length were sunk by such rogue waves between 1990 and 2004.

In 2005, while tracking the impacts of Hurricane Ivan, the U.S. Naval Research sensors in the Gulf of Mexico recorded a 91-foot (28-meter) wave. The largest estimated wave was 1,720 feet (524 meters) high and resulted from an underwater earthquake off the coast of southern Alaska in 1958.

Tides

The rotation of the Earth creates a gravitational force that holds the ocean water on the planetary surface. Tidal movement of the same water is caused by extra-planetary factors. The cyclical rising and falling of the surface of the ocean against the shores happens due to the intense forces of gravity being exerted by the moon and the sun on the Earth.

When the moon is new or full, the tides on Earth are at their highest and lowest levels, due to the planets' proximity to each other, in concert with the sun. The enhanced gravitational attraction draws water or pushes water away, depending on what side of the planet the moon is on. Conversely, when the moon is at its quarter phase, its gravitational forces are at right angles with that of the sun, resulting in lower tides.

The type of tide varies according to geographic location and water body dimensions. There are three basic tide patterns experienced across the oceans. Most locations host two similarly sized high tides and two low tides, also of similar size, in a twenty-four-hour period (called semidiurnal), cycling up or down every six hours and twelve minutes. Other areas only have one high and one low tide per day (called diurnal). Still other areas experience two high and two low tides (called a mixed tide), but one of each is noticeably larger than the other.

Ranges in tide intensity are caused by a number of factors, including wind strength, water depth, and land formations near the water. For example, the Bay of Fundy, off Canada's Atlantic coast, hosts some of the greatest tides, with a difference of 52 feet (16 meters) between low and high water. This large variation is caused by the unique funnel shape of the bay, which is sandwiched between the Canadian shores of New Brunswick and Nova Scotia.

Some of the smallest tidal ranges are just a few inches and are found in isolated smaller seas, such as the Sea of Japan and the Mediterranean, Baltic, and Caribbean Seas.

The Ocean's Power

The planetary ocean plays a central role in global temperature regulation, creating and regulating weather patterns, and balancing levels of carbon dioxide that have risen due to increased global man-made emissions. Despite its immense power, the ocean has limitations, and global climate change has impacted these vast saltwater bodies in several distinct ways.

Global Temperature Regulation

The World Ocean water body, with an estimated volume of 328 billion cubic miles (1.37 billion cubic kilometers), plays a central role in global temperature regulation. When the strong sun in summer months hits land, the heat is absorbed and radiated over several hours into the surrounding environment. When sunlight hits water, it either is reflected or absorbed. Because water is denser than air, this stored energy may be released days later in the same vicinity, or it may be moved elsewhere by currents. Conversely, cold climates chill ocean water, some of which is transported thousands of miles to warmer locations. This distribution mechanism helps maintain a global temperature balance, or equilibrium.

Weather Patterns

As the sun sends its rays into the ocean, the water absorbs energy during the summer months. During the winter, the water slowly releases the built-up energy. A similar process happens daily to a lesser degree, as daytime warming is followed by nighttime cooling by ambient air temperatures.

The surface waters of the ocean have a close relationship with the lower levels of the atmosphere. For example, onshore and offshore winds are influenced by ocean water.

Each day, as land near an ocean heats up under the rays of the sun, cool offshore ocean breezes blow across the land, helping to cool the region. During the night, the land cools to temperatures below that of the ambient air over the ocean, and the wind direction shifts, so that breezes blow from onshore—that is, from the land toward the water.

In temperate regions, warm air that blows across cold water creates fog. In tropical areas, heated water evaporates during the morning, leading to quick-forming thunderstorms in the afternoon and evening hours. In fact, these ocean-influenced weather patterns can result in severe weather when the water temperature gets high enough. This is especially common in the equatorial regions and the southwestern Pacific, mid-Atlantic, and southern Indian oceans. Warm waters (above 80 degrees Fahrenheit, or 27 degrees Celsius) result in high evaporation rates from ocean surfaces into the air, and such evaporation causes storms, which build up intensity as they feed on the energy of warm ocean waters.

The result can be a giant swirl of moisture-laden winds, known as spiraling condensation. More heat on the ocean's surface results in an increase in air density

HURRICANE CATEGORIES

Category	Wind Speed (Miles per Hour)	Ocean Surge (Feet Above Normal)	Impacts
One	74–95	4–5	Tree damage, pockets of coastal flooding, damage to older, weaker homes.
Two	96–110	6–8	Trees blown over, broad low-lying flooding, wider home damage.
Three	111–130	9–12	Large trees uprooted, extensive flooding, structural damage to many homes.
Four	131–155	13–18	All vegetation heavily impacted, extreme flooding, many homes destroyed.
Five	> 155	> 18	Complete destruction of impacted area. Only three have hit the United States since 1935.

Source: Saffir-Simpson Hurricane Scale, U.S. National Weather Service, 2009.

and corresponding wind speed. Spinning counterclockwise in the north and clockwise in the south, such a storm—which may be categorized as a tropical storm, cyclone, or hurricane—may stretch 600 miles (965 kilometers) in diameter and release more energy per day than seventy times the worldwide energy consumption.

The northern Pacific region, on average, generates up to twenty-seven tropical storms (up to 51 miles per hour, or 82 kilometers per hour), seventeen tropical cyclones (up to 94 miles per hour, or 151 kilometers per hour), and nine Category 3 hurricanes (142 miles per hour or more, or 228 kilometers per hour) per year. In the North Atlantic, by comparison, an average of eleven tropical storms, six tropical cyclones, and two Category 3 hurricanes are formed each hurricane season. Both of these northern seasons run from June to November.

Climate Change and Global Warming

As the planet's population grew from 978 million people in 1800 to 6.7 billion in 2009, its industrial economies generated considerable airborne pollution, mainly from factories, power plants, and vehicle emissions. This pollution has led to a phenomenon known as global climate change, which is happening due to an intense buildup of atmospheric "greenhouse" gases—such as carbon dioxide, methane, ozone, and nitrous oxide—and has been causing planetary warming.

Atmospheric carbon dioxide levels have risen 35 percent since the mid-eighteenth century, and average temperatures since 1996 have been the highest recorded in human history.

The eighteen largest glaciers in Norway and the largest in Europe are located on Svalbard, an archipelago of nine islands in the open Arctic Ocean north of mainland Norway. This once rare waterfall coming off Brasvell Glacier, located on one of the islands of Svalbard, has become commonplace due to the recent increase in air and water temperatures from global warming. An international assessment is under way to calculate the rate of loss of the glacier. *(Daisy Gilardini/Taxi/Getty Images)*

In 2007, the International Panel on Climate Change (IPCC), made up of members of the United Nations and the World Meteorological Organization, released a multi-year study on global warming. The scientific work of 600 authors from forty countries, titled *The Fourth Assessment on Climate Change 2007*, found: "Global atmospheric concentrations of carbon dioxide, ethane, and nitrous oxide have increased markedly as a result of human activities since 1750."

The concentration of carbon dioxide in the atmosphere in 2009 exceeded 387 parts per million. In 1800, it was less than 250 parts per million. Based on IPCC assessments, that level is climbing by roughly 2 parts per million per year and could surpass 500 parts per million by 2050. This change could result in an average global temperature increase of between 3.5 and 8 degrees Fahrenheit (-15.8 and -13.3 degrees Celsius) by the year 2100.

The oceans play a key role in balancing worldwide levels of atmospheric carbon dioxide. They absorb enormous quantities of carbon from the atmosphere and through decaying sea life, such as zooplankton, which ends up directly in the water. Oceans also help stabilize global temperatures by moving hot water to cold climates and cold currents to hot regions, thus helping to average temperatures to some degree worldwide.

Noticeable impacts to the oceans from climate change include an increase in ocean temperatures, melting of the polar ice, and general stress on ocean biota. As the air gets hotter, so do the oceans. This brings more intense, larger, more sudden, and longer-lasting storms.

In addition, ocean water absorbed up to 120 billion tons of carbon that was released into the environment by humans in the twentieth century. This additional carbon has saturated biotic life and resulted in an altered chemical makeup. Ocean water pH has been altered by 0.1, an almost 30 percent change since 1900. This phenomenon of ocean acidification pushes the normal pH range (7.8 to 8.5) downward and can impact all ocean life. Warmer ocean water and winds also have contributed to widespread polar ice melting. This has raised sea levels up to 8 inches (20 centimeters) since 1900. As more greenhouse gases are released by industrial powerhouses, such as China, Europe, India, and the United States, temperatures are forecasted to increase by as much as 4 degrees Fahrenheit (-15.6 degrees Celsius) during the first half of the twenty-first century. The International Panel on Climate Change has estimated that, by 2100, sea levels may rise another 20 inches (51 centimeters), causing extensive flooding in coastal and island regions.

The summer of 2009 saw two entire channels open at the North Pole, allowing the first commercial ship passage in more than 100 years from Canada to Denmark. The newly formed 1 million square miles (2.6 million square kilometers) of open water are expected to freeze again during the six months of winter, but thinner ice will mean more melting in the coming years, threatening coastal communities with widespread flooding from increased sea levels.

Selected Web Sites

British Oceanographic Data Centre: http://www.bodc.ac.uk.

Intergovernmental Oceanographic Commission: http://ioc-unesco.org.

International Ocean Institute: http://www.ioinst.org.

National Oceanic and Atmospheric Administration, Ocean Explorer: http://www.oceanexplorer.noaa.gov.

United Nations Atlas of the Oceans: http://www.oceansatlas.org.

Woods Hole Oceanographic Institution: http://www.whoi.edu.

Further Reading

Ballesta, Laurent, and Pierre Descamp. *Planet Ocean: Voyage to the Heart of the Marine Realm*. Washington, DC: National Geographic Society, 2007.

Cramer, Deborah. *Smithsonian Ocean: Our Water, Our World*. Washington, DC: Smithsonian Books, 2008.

Devey, Colin William, ed. *Oceans*. New York: Springer, forthcoming.

Dinwiddie, Robert, et al. *Ocean: The World's Last Wilderness Revealed*. London, UK: DK, 2006.

Earle, Sylvia A., and Linda K. Glover. *Ocean: An Illustrated Atlas*. Washington, DC: National Geographic Society, 2009.

Garrison, Tom. *Oceanography*. Belmont, CA: Brooks/Cole, 2007.

Intergovernmental Panel on Climate Change. *IPCC Fourth Assessment Report: Climate Change 2007 (AR4)*. Geneva, Switzerland, 2007.

Kolbert, Elizabeth. "The Darkening Sea." *The New Yorker*, November 20, 2006.

Pernetta, John. *Guide to the Oceans*. Buffalo, NY: Firefly, 2004.

Revkin, Andrew C. "Arctic Melt Unnerves the Experts." *The New York Times*, October 2, 2007.

Sverdrup, Keith, and E. Virginia Armbrust. *An Introduction to the World's Oceans*. New York: McGraw-Hill Higher Education, 2008.

2 | Humans and Oceans

Earth is the only planet in our solar system that contains a liquid ocean, an oxygen-based atmosphere, and the widespread presence of biotic life. The global ocean supports both aquatic and terrestrial upland life. Oceans' moderate temperatures have considerable bearing on weather, precipitation, and food growth on land, even affecting farmlands far inland. Without the ocean influence, far less rain would fall and food could not be grown as easily.

In the United States alone, the oceans annually support the transit of $700 billion in goods, supply $40 billion in natural gas and oil, provide $28 billion in seafood, and support an $11 billion tourism industry. However, humankind's attempt to explore, tame, and harvest the oceans has taken a toll on these vast bodies of water. Unsustainable fishing practices, habitat degradation, eutrophication (a decline in water quality due to excessive nutrients), and toxic pollution affect ocean health. Only in the last few decades have private industry and governments begun to work together to better manage and protect the oceans.

There are an estimated 215,000 documented species of animals, plants, and microorganisms living in saltwater habitats. More than 1 million species may exist, and an estimated 14 million in total have existed over the last 500 million years.

Ocean Ownership

Human ownership of oceans has been debated for centuries. The core question is: How far can a nation extend its legal boundaries from land into offshore waters?

Matthew Fontaine Maury (1806–1873), known as the father of modern oceanography, was a nineteenth-century scientist whose study of ocean currents helped ships find quicker routes across the oceans. Although the notion of a Northwest Passage—a sea route through the Arctic Ocean connecting the Atlantic and Pacific oceans—had been proposed by many explorers, Maury's 1855 *The Physical Geography of the Sea* helped prove its existence by discussing the migration patterns of whales that require the presence of open water for breathing. *(Hulton Archive/Stringer/ Getty Images)*

Throughout history, more than a dozen treaties have been drafted between various countries in an attempt to resolve the issue. Despite such structure being imposed over the years, continued disputes are common, especially when access to resources such as fishing grounds are at stake.

The first laws of the sea were written around 800 B.C.E. by the residents of a small Greek island called Rhodes in the Aegean Sea. This independent fishing community had a powerful commercial boat fleet and sought to apply standards to deal with shipping or other "on the water" issues. This three-part set of laws, eventually adopted and greatly expanded by the Romans, addressed naval issues, shipboard regulations, and maritime law in the Mediterranean Sea. Known as the Rhodian Code, it was observed by the Romans, Greeks, and other peoples for up to 1,000 years. One of the earliest provisions of the Rhodian Code set out to address property lost at sea:

> It is provided by the Rhodian Law that where merchandise is thrown overboard for the purpose of lightening a ship, what has been lost for the benefit of all must be made up by the contribution of all.

In 1602, the Dutch East India Company, a powerful government-sponsored trading enterprise that oversaw colonial activities in Asia, captured a Portuguese vessel in the straits of Malacca between the Pacific and Indian oceans. This action, which drew outrage and military threats from Portugal, was a single incident in a long list of conflicts concerning commerce on the open ocean. The Spanish fleets had claimed most of the Pacific Ocean and the Gulf of Mexico as theirs, while Portugal's armada had claimed portions of the Indian Ocean and much of the South Atlantic.

In 1608, a Dutch jurist named Hugo Grotius (acting anonymously because he was being paid by the Dutch East India Company) outlined a proposed guideline to allow for friendlier competition on the high seas. Grotius proposed a rule for how far out a country could claim ocean territory in his little book titled *Mare Liberum* (Free Seas). He described all of the oceans as international territory with the exception, labeled the "cannon shot" rule, being the distance that a nation's military cannons could fire from shore. This allowed for an approximately 3-mile (4.83-kilometer) extension of sovereignty from upland into ocean bodies. While written during a period of considerable conflict, *Mare Liberum* influenced future ocean laws and treaties.

At the end of World War I, in 1918, the United States took the unilateral step of asserting its sovereign right to navigate the ocean waters. This policy, called "freedom of the seas," was modified in 1945 to provide the United States with a 12-mile (19-kilometer) extension of its sovereign territory offshore. This caused a number of similar claims to be generated by other countries in the following decades, some extending up to 200 miles (322 kilometers) offshore in order to protect fishing grounds and other resources.

The primary modern international law governing the use and ownership of the oceans is the United Nations Convention on the Law of the Sea (UNCLOS). Written by the United Nations in 1958 and revised in 1982 and 1994, UNCLOS replaces multiple older treaties and outlines the rights and responsibilities of nations in their use of the world's oceans. The convention establishes guidelines for businesses, as well as for the management of marine natural resources and protection of the environment. This law outlines a 12-mile (19-kilometer) territorial area extending off the shores of any participating nation that will enforce regulatory issues such as fisheries management, environmental laws, and mineral rights. The convention also permits a 200-mile (322-kilometer) Exclusive Economic Zone intended to further protect local fishing fleets.

As of 2009, 159 countries had joined the convention; the United States, as well as twenty-one other countries, had signed but not ratified the treaty, citing economic and legal concerns. Since the original version of UNCLOS was written, about a dozen regional agreements between two or more nations have been created in order to settle disputes over fishing grounds or other maritime issues.

Exploring the Oceans

Evidence of open water travel dates back at least 15,000 years to when dugout canoes were carved and rafts were assembled to cross bodies of water. These small craft were not built to combat rough seas, though, and thus were limited to relatively short distances and coastal surroundings.

Larger ships and improved technology, such as canvas sails and directional measurements, resulted in an expansion in travel and a substantial increase in ocean knowledge. Nearly 5,000 years ago, Egyptian boats performed some of the first recorded open water voyages across the Mediterranean. By 660 B.C.E., ships had circumnavigated Africa. By 600 B.C.E., Babylonian ships had sailed from West Africa to Asia in search of new trade routes. Early world maps, such as the one drawn by a Greek merchant named Cosmas Indicopleustes around 550 C.E., showed a route to India; his map displayed a central landmass surrounded by a large ocean.

By 975 C.E., Vikings in northern Europe began extended sea journeys in their 100-foot (30-meter) "long ships" across the Atlantic Ocean. Many early attempts at ocean passage from Western Europe sought an improved route to Asia to increase trade.

Between 1400 and 1700, significant ship voyages resulted in visits to the farthest reaches of the ocean, opening up an era of unprecedented discovery. By 1475, improved astronomical measurement devices and charts developed by British, French, Spanish, and Portuguese fleets helped skilled pilots navigate accurately by sighting either stars or the moon. By measuring the angle of his position in relation to bodies such as the North Star, a pilot could calculate his ship's latitude (a position north or south of the equator).

Notable explorers of this period include Christopher Columbus, who sailed from Spain to the Bahamas in 1492; John Cabot, who set sail from England and discovered Newfoundland in 1497; Ferdinand Magellan, who departed Portugal in 1519 and died in the Philippines while his ship successfully circumnavigated the globe; Francis Drake, who departed England in 1577 and successfully sailed around the world; Abel Tasman, who completed two voyages from The Netherlands to the South Pacific in 1642 and 1644; and James Cook, an Englishman who began a two-year trip to circumnavigate the globe in 1768. As navigational tools improved and knowledge of the world expanded, numerous ocean voyages demonstrated man's determined attempt to reach new locations and map the world.

Although most Europeans felt that no life could survive more than a few hundred feet below the ocean surface, due to the darkness and high atmospheric pressure at great depths, the science of oceanography emerged in the 1800s. James Rennell, an English geographer who studied the ocean extensively, pub-

EARLY OCEANOGRAPHY ACCOMPLISHMENTS

Date	Name	Country/Region	Accomplishment
127 B.C.E.	Hipparchus	Greece	Invents the astrolabe for navigation, as well as a grid to calculate latitude and longitude.
1000 C.E.	Vikings	Scandinavia	Venture far from land into open oceans for food and exploration.
1419	Prince Henry	Portugal	Opens the first school for ocean navigation.
1537	Geradus Mercator	Germany	Creates the first map produced for navigation showing the world's curves on a flat map.
1760	John Harrison	England	Invents the chronometer, used to accurately determine a ship's position.
1769	Benjamin Franklin	United States	Produces a detailed chart of the Gulf Stream currents.
1818	John Ross	Scotland	Gathers sediment samples of deepwater areas while sailing to the Bering Strait.
1855	Matthew Maury	United States	Publishes *The Physical Geography of the Seas*, the first book of oceanography.
1868	Thomas Huxley	England	Studies deep-sea sediments and proposes that life began in the oceans.
1888	Multiple biologists	United States	Establish the Marine Biological Laboratory at Woods Hole, Massachusetts.

lished data he had collected over many years in his 1832 *An Investigation into the Currents of the Atlantic Ocean.*

Matthew Fontaine Maury, an American scientist who often is referred to in historical texts as the father of modern oceanography, published his groundbreaking *Physical Geography of the Sea* in 1855. His work documented wind and water currents and the variability of ocean life at depths below those at which scientists previously had thought life possible.

By 1872, the British government supported a scientific research vessel, the HMS *Challenger,* aboard which scientists traveled the globe, documenting ocean life and geography. This study proved that sea life existed down to 27,000 feet (8,230 meters) below the surface and that active volcanoes and other forces were reshaping the ocean landscape.

Challenges and Fears at Sea

Despite a substantial increase in ocean travel and exploration by the mid-1800s, the ocean remained a perilous and unforgiving place. Storms pounded ships at sea, and poor navigation practices left vessels far off course or smashed on reefs or rocky coasts. Over time, the knowledge of the World Ocean grew immensely; however, it came at the cost of thousands of lives.

Many societies worship ocean deities, including Greek gods such as Oceanus and Poseidon and the Roman god Neptune. The Inuit people of North America's polar regions described a world called Adlivun located beneath the seas. There, a deity named Sedna, or Arnakuagsak, is the caretaker of the sea animals; in Greenland, this goddess was believed to ensure that hunters reaped enough bounty from the seas. In Japan, Ebisu is a god of fisherman and good fortune.

Before the scientific revolution of the twentieth century, mythical creatures, or sea monsters, were described in legends around the world. Not unlike the perceived images of heaven or hell, depictions of the ocean were often vicious and described on a large scale. A Leviathan is described in the Bible's Old Testament (Psalm 74, Job 41, and Isaiah 27) as a large sea creature.

One Norwegian legend told of an enormous sea monster (up to 1.5 miles, or 2.4 kilometers, long) called the Kraken that was capable of engulfing a ship and pulling it to the bottom of the ocean. It used an inky black cloud to mystify seafarers before devouring them. This monster was later determined to be a giant squid (*Mesonychoteuthis hamiltoni*), a deep-ocean animal that can grow up to 46 feet (14 meters) long and is found primarily in the Pacific and Southern oceans.

There also are myths that portray the ocean as covering lost cities. The most famous is called Atlantis, described by the Greek philosopher Plato around 360 B.C.E. and thought to be a Greek island in the Atlantic Ocean, just west of Spain. Plato's account was incomplete and motivated a number of stories, searches, and claims throughout history. In 2001, a remarkable actual lost city was discovered by oceanographers off the western coast of India. Sitting in 120-foot-deep (37-meter-deep) waters, the 5-mile (8-kilometer) north to south by 2-mile (3.2-kilometer) east to west village was a thriving community around 9,500 years ago. After the end of the most recent ice age, rising sea levels submerged the entire settlement.

The complete disappearance of a ship at sea was common up until 1850. Thereafter, with the development of radio communications and, later, satellite tracking, many threatened ships could better navigate or be located and rescued.

A shipwreck refers to an event where a ship runs aground and is severely damaged by waves or rocks onshore. The United Nations estimated in 2006 that 3 million shipwrecks cover the floors of the oceans. Two of the more well-known modern shipwrecks include the 882-foot (269-meter) RMS *Titanic* ocean liner, lost in 1912 in the North Atlantic Ocean, resulting in more than 1,500 deaths, and the 509-foot (155-meter) cruise ship MS *Estonia*, which sunk with 852 passengers while crossing the Baltic Sea in 1994.

Despite advanced navigation and weather forecasting, ocean conditions still claim up to a dozen ships a year. For example, on November 23, 2007, the 246-foot (75-meter) cruise liner MS *Explorer* struck an iceberg off the Antarctic Peninsula and sunk to a depth of 3,707 feet (1,130 meters) within a few hours. All 154 people on board were rescued by nearby ships.

Ships and Technology Advance

After 1700, the increase in trade and economic activity around the World Ocean brought steady improvements in ship and instrument designs. For example, bigger three-mast ships were built to carry a larger crew with more gear and food for lengthier travel than previous vessels. Navigation methods and trip planning also became more reliable. In addition to an increase in trade, these developments increased the amount of European colonization around the world.

The earliest version of a sextant was developed in 1731 by Englishman John Hadley (1682–1744), and it became an important tool for maritime navigation. Due to the fact that a sextant is not dependent on electricity or other modern technology, it still is considered a practical backup tool to determine a ship's latitude. (© Bernard Maurin/Fotolia)

One significant technological breakthrough was the invention of the sextant. This handheld device allowed the measurement of the position of the moon, the sun, or other celestial objects to determine a ship's latitude (the angle of the object would reveal the ship's distance from the equator). Designed by the Englishman John Hadley in 1731, the sextant replaced a number of less reliable navigational instruments.

Another challenge was determining the longitude of ships at sea. While the latitude could be measured with a sextant by calculating a horizontal location (north to south) in relation to the equator, a measurement of longitude (vertical position from east to west) in relation to the prime meridian (a line that runs from the North Pole to the South Pole) was much more challenging. In 1760, John Harrison, a clockmaker from Lincolnshire, England, built a special ship-based marine chronometer to provide an accurate time measurement device that worked well on rough seas. When Harrison's watch finished its first trans-Atlantic journey in 1761, it had only lost five seconds. With such a reliable clock, a mariner could calculate longitude by knowing the local time from a given location on land.

The next significant seafaring technology was the use of radio waves (called radar, which today stands for "radio detection and ranging") to send signals between land and ships at sea. The U.S. Navy began using ship-to-shore radar in 1902. A broad campaign to use radar signals globally resulted in the construction of hundreds of steel towers on islands and along shores. In the 1970s, the use of radio waves for navigation began to be replaced by satellite-based communications and the global positioning system (GPS), which consists of a network of up to thirty-two orbiting satellites. Radar still is used by some vessels to detect objects in conditions of low visibility.

Population Growth Causes Ocean Pollution

The world population increased by three and a half times between 1500 and 1900, from 450 million to 1.6 billion. From 1900 to 2009, it grew fourfold, from 1.6 billion to 6.7 billion. More people and larger economies have negatively impacted the oceans by fishing, mining, dumping land and ship waste, and creating unprecedented upland pollution (via air, land, and water pathways). The increased number of people reliant on commerce and resource consumption worldwide, particularly during the twentieth century, has resulted in considerable long-term problems for the oceans. The majority of ocean pollution (up to 80 percent) is from land-based human activities, such as storm water runoff from roads and impervious surfaces. Community infrastructure, including streets, parking lots, and rooftops, sends road salts, oil, fertilizers, garbage, human waste, and toxic chemicals into rivers and bays, and directly into oceans.

GLOBAL CARBON MOVEMENT PATTERNS
(Petagrams of Carbon/Year)

Location	Carbon Input	Carbon Output	Total Carbon Held
Land	120	119	2,000
Atmosphere	222.5	220	760
Fossil fuels	0	6.5	5,000
Oceans	100	97	38,800

Source: Richard Houghton. *Balancing the Global Carbon Budget*. Falmouth, MA: The Woods Hole Research Center, 2007.

In addition to problems caused by toxic pollution, such as untreated human sewage being discharged into waterways, extra nitrogen and phosphorus from lawns, septic systems, and farms feed large algae blooms in coastal ocean bodies. This has led to eutrophication, a condition where plant blooms use the available oxygen in the water, leaving less oxygen for other marine life and resulting in the loss of fish and other sea creatures.

Beneath the ocean floor, large oil and natural gas deposits from decaying organic matter have formed over the last 2 billion years in pockets of rock. These resources first were sought on land, but underwater exploration efforts since the turn of the twentieth century found considerable oil and gas deposits offshore. The first Pacific Ocean oil wells were drilled in 1896 off the coast of Santa Barbara, California. Since then, two-thirds of the planetary continental shelf has been found to hold rich deposits. By 1975, efforts to locate and retrieve oil and gas had moved offshore into deeper waters, with seventy-seven U.S. well stations in waters up to 650 feet (198 meters) deep.

The North Sea, which is bounded by the Orkney Islands and east coasts of England and Scotland to the west and the northern and central European mainland to the east and south, hosts large oil deposits, first located in 1960. The British and Norwegian sections of the sea contain most of the large oil reserves. By 1989, 149 large drilling platforms had been constructed, and by 1999, 5,000 miles (8,045 kilometers) of underwater pipeline had been laid to distribute 6 million barrels per day.

As the quality and accessibility of concentrated metals mined on land decreased through the twentieth century, the mining industry began searching the oceans for high-grade copper, iron, lead, and nickel. The bottom of the oceans contains, in places, an abundance of these materials. Although limited technology and high costs prohibited large-scale excavation until the late twentieth and early twenty-first centuries, new processing boats, designed to bring up and separate rocky materials on-site, have made marine mining possible. In 1994, the United Nations formed the International Seabed Authority, an autonomous international organization, to regulate this new industry.

Household, chemical, dredged, industrial, medical, and nuclear waste has been commonly dumped at sea. In 1972, the United States and the United Nations imposed new rules intended to regulate such practices. Since that time, seventy-eight countries have agreed to comply with a United Nations ocean dumping law, which was updated in 1996.

The ocean, however, still is used as an illegal dumping ground for millions of tons of garbage every year. Little enforcement, low penalties, and the difficulty of catching a polluter in the act are common challenges. As dumped waste often is preserved in a watery state, it coats the bottom of the ocean with a toxic brew that harms local marine life; it also can wash ashore during storms, threatening wildlife and coating beaches.

Ocean vessels historically were unregulated and released chemicals, fuel, garbage, human waste, and oil into the seas. The United Nations International Convention for the Prevention of Pollution from Ships (the MARPOL Protocol) was created in 1973 to reduce boat dumping. Updated through the years and adopted by 138 countries as of 2009, the convention regulates ships by requiring certain equipment, as well as periodic inspections. Although these improvements improve efficiency and storage, the convention does not prevent certain releases at sea. Instead, it permits such releases at a distance from shore, including oil (50 miles, or 80 kilometers, out to sea), untreated garbage (12 miles, or 19 kilometers), and untreated sewage (12 miles, or 19 kilometers).

Air pollution was not addressed as a threat to oceans until the late twentieth century, when contamination of the atmosphere over the seas by airborne pollutants became a worldwide issue. Amendments to MARPOL in 2005 set air pollution limits.

Ocean Regulation and Scientific Research

Since the 1970s, governments and private industry have worked to protect and study the oceans. Local and international laws have begun to address pollution issues by limiting dumping. While boat access into harbors and waterways is important for efficient commerce, the practice of dredging (removing sediments where shallow waters impede travel) can harm marine life. New limits to coastal dredging practices have better protected fish and clam populations. There also has been a proliferation of large marine scientific research and educational facilities (now up to thirty worldwide) that have brought about a greater understanding of ocean dynamics, biology, and ecology than ever before.

Although the United Nations has orchestrated key treaties to advance broader ocean protection efforts, such as the Law of the Sea Convention and the International Convention for the Prevention of Pollution from Ships, regional agreements between two or more countries also have had significant impact.

For example, the Agreement on the Organization for Indian Ocean Marine Affairs was created in 1990 to increase ocean awareness, encourage member countries to share information on ocean systems, and enhance international dialogue on ocean issues. As of 2009, thirty-eight nations that border the Indian Ocean participated in the treaty.

When continental resources such as wetlands, wildlife, forests, and fresh-water have been overused, government oversight has been applied to better manage them for long-term, sustainable use. This approach was not used with the ocean environment until the 1980s, when a number of regional agreements acknowledged overutilization and established short- and long-term goals.

The Agreement Between the Government of Canada and the Government of the United States of America on Pacific Hake/Whiting (commonly called the Pacific Hake/Whiting Agreement) allows for recovery of Pacific whiting (or hake, *Merluccius productus*) populations by setting catch limits for each nation. Signed in 2003, the parties agreed that U.S. fishermen would be allocated 74 percent of the fish off the West Coast of the United States and British Columbia; Canadian fishermen would be allocated 26 percent. The regional fishing industry for Pacific whiting provides more than $25 million per year in total revenue.

In the early twentieth century, there were only a handful of private organiza-tions dedicated to the scientific research of oceans. One of them was the Woods Hole Oceanographic Institution, established in 1930 in the Cape Cod village of Woods Hole, Massachusetts. It grew from a small group of scientists to 148 scientists and 763 staff with a $158 million annual research budget in 2008. The organization has three research vessels, ranging between 177 and 274 feet (54 and 84 meters) in length, and several deep-diving submarines, including the *Alvin*, which has hosted 12,000 researchers on hundreds of diving research trips since 1964. The institution performs a broad range of studies in subjects such as fish health, pollution impacts, saltwater chemistry, and ocean floor mapping.

The Woods Hole model of a marine research organization dedicated to ocean studies and public education has been implemented successfully elsewhere. Other such marine organizations include the Ocean Research Institute at the University of Tokyo, founded in 1962 in Tokyo, Japan; the Alfred Wegener Institute for Polar and Marine Research, established in 1980 in Bremerhaven, Germany; and the Australian Institute of Marine Science, founded in 1972 in Townsville, Australia.

Selected Web Sites

Alfred Wegener Institute for Polar and Marine Research: http://www.awi.de/en/home.

Australian Institute of Marine Science: http://www.aims.gov.au.

National Maritime Museum: http://www.nmm.ac.uk.

NEPTUNE: http://www.neptune.washington.edu.

Ocean Research Institute, University of Tokyo: http://www.ori.u-tokyo.ac.jp/index_e.html.

United Nations International Maritime Organization: http://www.imo.org.

United Nations Oceans and Law of the Sea: http://www.un.org/Depts/los/.

Woods Hole Oceanographic Institution: http://www.whoi.edu.

Further Reading

Buschmann, Rainer. *Oceans in World History*. New York: McGraw-Hill, 2006.

Byatt, Andrew, Alastair Fothergill, and Martha Holmes. *The Blue Planet*. New York: DK, 2001.

Langewiesche, William. *The Outlaw Sea*. New York: North Point, 2004.

Maury, Matthew Fontaine. *Physical Geography of the Sea*. Boston: Adamant Media, 2001.

Roberts, Callum. *The Unnatural History of the Sea*. Washington, DC: Island, 2007.

Rozwadowski, Helen M. *Fathoming the Ocean: The Discovery and Exploration of the Deep Sea*. Cambridge, MA: Belknap Press, 2005.

Steinberg, Philip E. *The Social Construction of the Ocean*. New York: Cambridge University Press, 2001.

Weir, Gary E. *An Ocean in Common*. College Station: Texas A&M University Press, 2001.

3 | Ocean Locations

The existence of saltwater oceans across nearly three-quarters of the Earth's surface resulted in the name "The Blue Planet" in 1968 when the United States' Apollo 8 space mission returned from its moon orbit with a series of remarkable color photographs. Taken from 238,000 miles (382,942 kilometers) away, the images gave viewers a sense of humankind's limited place in the universe, as well as the dominant presence of the colorful oceans.

Despite the fact that a single World Ocean exists, governments have long used regional names to describe ocean bodies, with five in particular that are the largest in area and depth.

Atlantic Ocean

Named after the Greek mythic titan Atlas, a deity described as having superhuman strength, the Atlantic Ocean covers up to one-fifth of the world's surface area. Its S shape spans 33.5 million square miles (87 million square kilometers), making it the second-largest ocean on the planet. It has an average depth of 11,828 feet (3,605 meters). The deepest mapped area is 28,233 feet (8,605 meters) and is located inside the Puerto Rico Trench, an oceanic trench on the boundary between the Caribbean Sea and the Atlantic Ocean.

The Atlantic Ocean is divided into two distinct areas, north and south; these areas meet near the equator where the continents of South America and Africa are closest together (1,700 miles, or 2,735 kilometers, apart). It is bordered on its southern edge by the Southern Ocean and on its northern edge by the Arctic Ocean. Although the Atlantic Ocean is fed by the most freshwater runoff of any of the world's oceans—as it is adjacent to the large continents of Europe, North and South America, and Africa—it also is the saltiest (ranging from 3.3 to 3.7 percent salinity), due to currents, evaporation rates, and its flow through tropical areas.

The Atlantic Ocean formed 200 million years ago during the Cretaceous Period, first as an opening between South America and Africa and then, 100 million years ago, as an opening between North America and Europe. The central geological feature of the Atlantic is a massive, 8,800-mile-long (14,159-kilometer-long) volcanic ridge that runs from Iceland in the north to the Antarctic Basin in the south. Known as the Mid-Atlantic Ridge, it measures up to 1,000 miles (1,609 kilometers) at its widest point and rises tens of thousands of feet from the sea floor to form islands in several places, such as the Azores (930 miles, or 1,496 kilometers, west of Portugal) and Tristan da Cunha (1,750 miles, or 2,816 kilometers, from South Africa). The ridge is actively changing; it widens up to 2 inches (5 centimeters) each year, expanding the Atlantic Ocean through the release of new lava that turns into seabed. As these undersea volcanoes erupt, the new

This image shows the vastness of the World Ocean. Taken in 1968 by U.S. astronaut Bill Anders during the Apollo 8 space mission (the first time a manned spacecraft traveled into lunar orbit), it is the first such photograph of the Earth by a human. *(NASA/Time & Life Pictures/Getty Images)*

OCEAN FACTS

	Atlantic Ocean	Pacific Ocean	Arctic Ocean	Indian Ocean	Southern Ocean
Size (million square miles)	33.5	65.6	5.44	26.9	7.8
Length (miles)	8,774	8,637	3,107	41,300	13,360
Maximum Width (miles)	4,909	11,185	1,988	6,338	1,678
Maximum Depth (feet)	28,233 (Puerto Rico Trench)	36,201 (Mariana Trench)	18,456 (Near North Pole)	24,460 (Java Trench)	23,736 (South Sandwich Trench)
Primary Fish	Cod, grouper, hake, herring, mackerel, menhaden, tuna	Cod, hake, mackerel, pollock, salmon, sole, sardine, tuna	Arctic cod, arctic greyling, whitefish	Anchovy, flounder, grouper, herring, shark	Cod, herring, krill, squid
Deep Ocean Floor (greater than 6,000 feet; percent)	38	43	60*	49	50
Volcanoes and Trenches (percent)	2.8	5.4	1 or less*	6	1
Slopes (percent)	27.9	15.7	38*	15	29
Ridges (percent)	31.2	35.9	1 or less*	30	20

*Estimated figures.

material pushes the existing adjacent tectonic plates west toward North and South America and east toward Europe and Africa.

The bottom of the Atlantic Ocean features a dozen large basins called abyssal plains. These locations are noticeably flat and range between 7,000 and 18,000 feet (2,134 and 5,486 meters) deep. Often compared to a barren, windswept desert landscape, abyssal plains are characterized by mostly level terrain, total darkness, enormous water pressure, and some of the coldest waters on the planet. A jutting volcano called a seamount (also called a ridge) occasionally may divide these plains.

Additionally, the abyssal areas host fewer species of fish than other ocean areas; many species that are found here must swim upward to feed in shallower waters. To live at these high-pressure, cold depths, a number of species have evolved to have limited motion and have developed a slower metabolism. The species generally feed on the slow drift of organic materials—made up of dead fish, algae, plankton, and other sea life and waste streams—that sinks to deeper waters, eating this material as it accumulates on the ocean floor.

The Atlantic Ocean also features some of the richest marine environments on the planet. When deepwater currents collide with shallower coastal waters, large amounts of nutrients such as calcium, nitrogen, and phosphorus from the ocean bottoms are delivered to the surface as food for the smallest creatures, such as microscopic diatoms. Plankton consume these diatoms, small fish eat the plankton, and the small fish are then eaten by bigger fish. These intersections of deep and shallow water result in a diversity of plankton, fish, shellfish, mammals, and marine bird species.

Upwelling locations in the Atlantic Ocean include the Grand Banks, southeast of Newfoundland on the North American Continental Shelf; Georges Bank, between Cape Cod, Massachusetts, and Cape Sable Island, Nova Scotia; Hatton Bank in the Iceland Basin; the Bahama Banks of the Bahama Archipelago in the Caribbean; and the Falkland Banks, near Argentina.

Pacific Ocean

In 1520, Portuguese explorer Ferdinand Magellan named the largest ocean *Mare Pacificum,* meaning "peaceful sea" in Latin, when he encountered pleasant stillness after traveling for the first time through the turbulent straits (later named after him) that lead from the Atlantic Ocean to the Pacific Ocean at the southern tip of South America. The Pacific Ocean covers up to one-third of the Earth's surface, with an area of 65 million square miles (168 million square kilometers). It is the deepest ocean, with an average depth of 13,127 feet (4,001 meters). Its deepest known point is at 36,201 feet (11,034 meters) in the Mariana Trench, to the east of the Mariana Islands (south of Japan and north of New Guinea). The Pacific Ocean runs from the Arctic Ocean (at the intersection of Russia and the United States) in the north to the Southern Ocean in the south, ending at 60° latitude.

The collision and constant movement of tectonic plates transformed the historic Panthalassa into the Pacific Ocean of today. Although it is the largest ocean, the Pacific is shrinking by up to 4 inches (10 centimeters) per year. This is due to several tectonic continental plates (Asian, Australian, North American, and South American) pressing up against the ocean plates, forcing the denser ocean plates downward into trenches. Once pressed down, these steep valleys are forced into subduction zones, where the ocean floor either is driven into the magma that sits under the crust or is pushed under another continental plate and eventually crushed and melted.

The northern Pacific Ocean hosts a horseshoe-like geologic formation called the Ring of Fire that is made up of undersea and island volcanoes and ocean trenches. This ring, running for 24,850 miles (39,984 kilometers) from New Zealand to South America, is the source of frequent disturbances, including earthquakes, tsunamis, volcanic eruptions, and ocean floor subductions. The Ring of Fire includes an estimated 450 volcanoes, thousands of volcanic vents, and 25,000 islands. It hosts 90 percent of the world's earthquakes. The source of this activity is the Pacific tectonic plate colliding into a dozen others, including Nazca, Cocos, South and North American, and Juan de Fuca.

Nearly 40 percent of the fish in the World Ocean live in the Pacific Ocean. This includes several dozen groups encompassing up to 20,000 different species, most of which are bony-type fish (exceptions include rays and sharks). Some of the most common larger Pacific commercial fish are six species of salmon (*Oncorhynchus sp.*), three species of tuna (*Thunnus sp.*), and two species of bass (*Stereolepis sp.*). Small commercial fish include the Pacific herring (*Clupea pallasii*), which is primarily caught to make by-products such as fish meal and food additives.

Until the middle of the twentieth century, herring were so abundant that they made up 30 percent of the weight of the total global annual catch of all fish. Overfishing, which resulted in a collapse of the eastern Pacific population in 1993, has harmed herring stocks, as well as other larger fish (tuna, salmon, and bass) that feed on the herring.

The ocean bounty allowed for the development of a large Pacific fishing fleet, which operates mostly along coastal regions in the shallower waters of continental shelves. This fleet of over 1 million boats brings in 60 percent of the global fish catch each year, including 5 billion pounds (2.3 billion kilograms) of tuna alone. The largest fishing ports are spread out across the ocean and include Callao, Peru; Hong Kong, China; Klang, Malaysia; Sydney, Australia; Tokyo, Japan; and the American ports of Los Angeles, California, and Seattle, Washington.

Arctic Ocean

The smallest of the major ocean bodies, the Arctic Ocean is located in the polar regions of the northern hemisphere. It is framed by a nearly complete circle of land made up of portions of Asia, Europe, Greenland, North America, and Russia. The 5.4-million-square-mile (14-million-square-kilometer) Arctic Ocean also is the shallowest ocean, with an average depth of 4,690 feet (1,430 meters). The deepest point, at 18,456 feet (5,625 meters), is near the North Pole at the center of the ocean.

Hosting extremely cold winters marked by months of total darkness, this arctic region maintains a prominent ice cap (a literal ice blanket composed of freshwater) that spreads up to 5 million square miles (13 million square kilometers) across the ocean. The ice at the North Pole is up to 164 feet (50 meters) deep in some places. These arctic conditions prevent animal or human habitation in the areas closest to the pole during the winter months.

Summers, especially as influenced by the impacts of climate change, have experienced seasonal melting of up to half of the pack ice. Scientists predict the complete disappearance of summer ice before the end of the twenty-first century. For the first time in history, large commercial vessels can ply the open channels of the Arctic Ocean in the summer.

The Arctic Ocean is one of the least-explored areas of the world due to its relative inaccessibility. It was not until 1958 that the U.S. submarine *Nautilus* was able to surface through the summer ice at the North Pole to prove that only ice covers the surface of the ocean.

The decades of the cold war between the United States and the Union of Soviet Socialist Republics (USSR) led to heavy submarine traffic under the ice and the first efforts to map the ocean floor. Sonar surveys performed in 1990 gathered the first depth measurements and detailed images of the floor.

One distinct feature of the floor is that more than 50 percent of it consists of a shallow continental shelf, especially where it connects to portions of Canada and Russia. At a depth of less than 600 feet (183 meters), the shelf extends up to 1,000 miles (1,609 kilometers) toward the pole.

The Arctic Ocean is so shallow in places that some oceanographers consider it a sea, although a sea typically is defined by its relationship to an adjacent landmass and not by its depth. Despite this debated terminology, the Arctic Ocean is growing in size along the Arctic Mid-Ocean Ridge, located between Greenland and Siberia, as well as along three smaller ridges.

Although its ice coverage changes by the season, the arctic polar ice cap grows each winter to approximately 5.7 million square miles (14.8 million square kilometers) and covers almost the entire ocean area plus portions of Greenland and the Bering Strait (located between Cape Dezhnev, Chukotka, Russia, and Cape Prince of Wales, Alaska). This cap, while up to 150 feet (46 meters) thick in places, contains only 3 to 4 feet (0.9 to 1.2 meters) of new ice each year. Some of the new ice on the outer portion of the cap historically has melted in the summer months from June to September; this melting has increased substantially due to the impacts of climate change.

The Arctic Ocean has limited inflow of salt water from other oceans—nearly 80 percent enters the Arctic from the Atlantic Ocean next to Greenland and Norway. This limited water inflow, along with heavy inflow from freshwater land runoff, plays a key role in the formation of the polar ice cap.

The extremely cold conditions in the polar region—temperatures stay below 32 degrees Fahrenheit (0 degrees Celsius) year-round on the ice cap—make life for biotic creatures very difficult. The Arctic Ocean, however, hosts a handful of mammals that have evolved to thrive in a harsh ocean environment where ice functions like land.

These include polar bears (*Ursus maritimus*), ringed seals (*Pusa hispida*), and walrus (*Odobenus rosmarus*). The web of life in the Arctic Ocean begins with plankton growing in the upper layers of salt water near the floating ice. These plant and animal species are consumed by small fish that are then eaten by species such as Arctic cod (*Boreogadus saida*). The cod are hunted by three species of seals that are, in turn, eaten by polar bears. The 1,300-pound (590-kilogram) polar bear, which is camouflaged by its white coat, is semi-aquatic, and it spends much of its life on the ice pack as a consummate predator.

Indian Ocean

Covering up to 20 percent of the Earth's surface, the Indian Ocean is the third-largest major ocean. It is bordered by Africa to the west, India and Asia to the north, Australia to the east, and the Southern Ocean to the south. It spans 26.9 million square miles (69.7 million square kilometers) and has an average depth of 12,645 feet (3,854 meters). The deepest location is the 24,442-foot (7,450-meter) Java Trench, which lies approximately 186 miles (299 kilometers) off the coasts of Java and Sumatra.

The Indian Ocean formed after Gondwanaland, a conglomeration of continents including Africa and Australia, broke apart 130 million years ago. India sat in the middle of this new ocean until 75 million years ago, when it began to travel northward into the Asian tectonic plate. Eventually, the Indian continent collided with Asia and, 35 million years ago, began to form the Himalayan Mountains.

Since then, the Indian Ocean has continued to grow. Two of the world's largest rivers, the Brahmaputra of southern Tibet and the Indus in Pakistan, drain into this ocean and deposit enormous amounts of sediment up to 1,200 miles (1,931 kilometers) into the open waters.

Along its floor, the Indian Ocean has an inverted Y-shaped formation made of undersea volcanic ridges. One section, the Ninety East Ridge, runs from eastern India in the Bay of Bengal 1,700 miles (2,735 kilometers) southward. This prominent underwater mountain range sits 5,500 feet (1,676 meters) below sea level due to erosion and plate tectonics, but it once stood above the surface waters. The ridge contains portions of the African, Indonesian, and Australian continents on its crest. Roughly 70 million years ago, as the Indian continent traveled northward to collide with Asia, its movement against the edge of the Ninety East Ridge left a noticeable wake of crushed and impacted underwater rocky terrain, visible by sonar surveys.

Warm waters help create monsoons (powerful storms similar to hurricanes) in the Indian Ocean during two periods each year. The first, from November to April, features steady winds and currents that flow generally from east to west. Ship travel from India to Africa is easier during this time, and relatively cooler weather means calmer coastal conditions. The second monsoon season, from May to October, is the southwest monsoon season. The winds and currents shift dramatically and instead travel from west to east. Increased air temperatures result in much warmer waters that create numerous ocean and coastal storms.

Many low-lying areas, such as coastal India, Sumatra, and Thailand, are particularly susceptible to damage from flooding. Winds in excess of 150 miles per hour (241 kilometers per hour) may occur up to a dozen times a season, and ocean waves cresting above 40 feet (12 meters) are common.

The Indian Ocean features unique islands, called atolls, formed by oceanic volcanoes that eroded to form ring-shaped landmasses, each surrounding a saltwater central lagoon. These atolls, whose name means "place" in the native language of the Maldive Islands peoples living southwest of India, support the formation of underwater coral reefs that host a diverse community of sea life. The shallow, warm waters of an atoll attract coral organisms that slowly turn nutrient-poor waters into a viable habitat for up to 4,000 species of tropical fish. Up to 300 atolls exist around the world, with the majority being in the Pacific Ocean and Indian Ocean.

Southern Ocean

It was not until the year 2000 that the international community gave a new name to this existing ocean body. What today is termed the Southern Ocean has had several names, including the Antarctic Ocean and South Polar Ocean. Scientists have formally acknowledged it as a fifth distinct ocean.

Portions of the Atlantic, Pacific, and Indian oceans were carved away at the 60° south latitude line to create a circular body of water that surrounds the Antarctic continent, forming the fourth-largest ocean. Spanning 7.8 million square miles (20.2 million square kilometers), the Southern Ocean has an average depth of 14,450 feet (4,404 meters), with its deepest point at 23,736 feet (7,235 meters) in the South Sandwich Trench area, east of the tip of South America. Because of its location at the polar south region, up to 1.1 million square miles (2.8 million square kilometers) of the ocean freeze around Antarctica each winter.

At the heart of the Southern Ocean is the coldest continent on the planet, Antarctica, where winter features no daylight and temperatures that reach -85 degrees Fahrenheit (-65 degrees Celsius). This frozen continent is completely covered by snow and ice (averaging 1 mile, or 1.6 kilometers, in thickness) and has

The largest population of Gentoo penguins (*Pygoscelis papua*), distinguishable by their bright red-orange bills and white patches behind their eyes, is on the coastal areas of the Antarctic Peninsula. A thick layer of blubber and network of waterproof plumage enables these birds to withstand the incredibly cold waters and harsh Antarctic weather. (© lfstewart /Fotolia)

a very narrow continental shelf with ocean water reaching up to 1,600 feet (488 meters) deep as far as 100 miles (161 kilometers) from land.

In fact, Antarctica holds 90 percent of the Earth's ice, which is 70 percent of the worldwide freshwater. This snow and ice play a central role in the planetwide thermoregulation of air temperatures by reflecting a high amount of sunlight back into space. The continent's inhospitable features—no trees, little exposed rock, and powerful multi-day storms with whiteouts—provide poor conditions for animal survival, and thus the edge of the landmass, where the ocean meets the land, is where almost all wildlife lives.

The area defined as the edge of the Southern Ocean, at 60° latitude, is the only place on the planet where no land exists for the entire circumference of the globe. It is here that the open ocean, which hosts waters exceeding 20,000 feet (6,096 meters) deep, features considerable waves that could, feasibly, travel in a complete circle around the planet without being blocked by land.

Ship captains, even those with very large vessels, have for years called this area the "screaming sixties," referring to the latitude, and describing the continuous howl of unimpeded gale-like winds. Many of the world's largest waves—often in excess of 50 feet (15 meters) high—have been witnessed in the Southern Ocean.

The combination of deep, cold water and powerful currents supports a bounty of life in Southern Ocean waters, including an estimated yearly production of 610 million tons of phytoplankton. These plants feed schools of small fish that support a number of seabirds, seals, and whales.

The largest penguin species, the emperor (*Aptenodytes forsteri*), spends much of its life feeding in the ocean; it breeds a few miles inland on Antarctica during the brutal winter months. Feeding on tiny shrimp called krill, small fish, and squid, the penguins grow up to 4 feet (1.2 meters) tall and weigh nearly 100 pounds (45 kilograms).

The most populous seal in the Southern Ocean is the crabeater (*Lobodon carcinophagus*), which survives by filtering water through its lobed teeth, thus capturing mouthfuls of tiny shrimp (rather than crabs as its name suggests). With a global population of up to 50 million, this seal is the second-most-populous large mammal on the planet after humans.

Selected Web Sites

National Institute of Oceanography, India: http://www.nio.org.

National Oceanography Centre, University of Southampton: http://www.noc. soton.ac.uk.

Monterey Bay Aquarium Research Institute: http://www.mbari.org.

University of Otago, Marine Science: http://www.otago.ac.nz/marinescience/.

Further Reading

Ballesta, Laurent, and Pierre Descamp. *Planet Ocean: Voyage to the Heart of the Marine Realm*. Washington, DC: National Geographic Society, 2007.

Dinwiddie, Robert, et al. *Ocean: The World's Last Wilderness Revealed*. London, UK: DK, 2006.

Earle, Sylvia A., and Linda K. Glover. *The National Geographic Atlas of the Ocean*. Washington, DC: National Geographic Society, 2001.

Ellis, Richard. *The Empty Ocean: Plundering the World's Marine Life*. Washington, DC: Island, 2003.

Hutchinson, Stephen. *Oceans: A Visual Guide*. Buffalo, NY: Firefly, 2005.

Pearson, Michael. *The Indian Ocean*. New York: Routledge, 2007.

Pernetta, John. *Guide to the Oceans*. Buffalo, NY: Firefly, 2004.

Rose, Paul, and Anne Laking. *Oceans*. Berkeley: University of California Press, 2009.

Soper, Tony. *The Arctic Ocean*. Guildford, CT: Globe Pequot, 2001.

Sverdrup, Keith, and E. Virginia Armbrust. *An Introduction to the World's Oceans*. New York: McGraw-Hill Higher Education, 2008.

Trujillo, Alan P. *Essentials of Oceanography*. Upper Saddle River, NJ: Prentice Hall, 2007.

OCEANS OF THE WORLD
CASE STUDIES

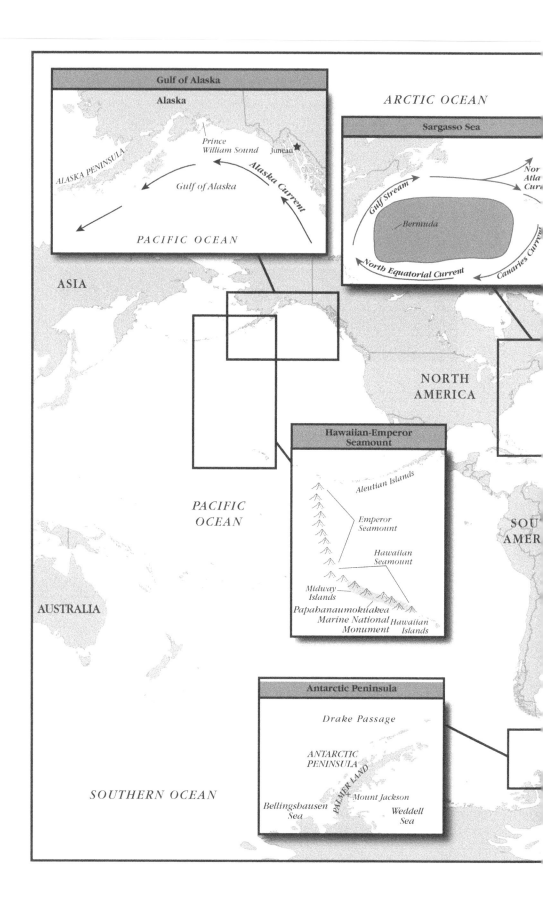

Gulf of Alaska

Alaska

Prince William Sound Juneau★

ALASKA PENINSULA

Alaska Current

Gulf of Alaska

PACIFIC OCEAN

Sargasso Sea

ARCTIC OCEAN

Gulf Stream

Nor Atla Cur

Bermuda

North Equatorial Current

Canaries Current

ASIA

NORTH AMERICA

PACIFIC OCEAN

Hawaiian-Emperor Seamount

Aleutian Islands

Emperor Seamount

Hawaiian Seamount

Midway Islands

Papahanaumokuakea Marine National Monument

Hawaiian Islands

SOU AMER

AUSTRALIA

Antarctic Peninsula

Drake Passage

ANTARCTIC PENINSULA

PALMER LAND

+ *Mount Jackson*

Bellingshausen Sea

Weddell Sea

SOUTHERN OCEAN

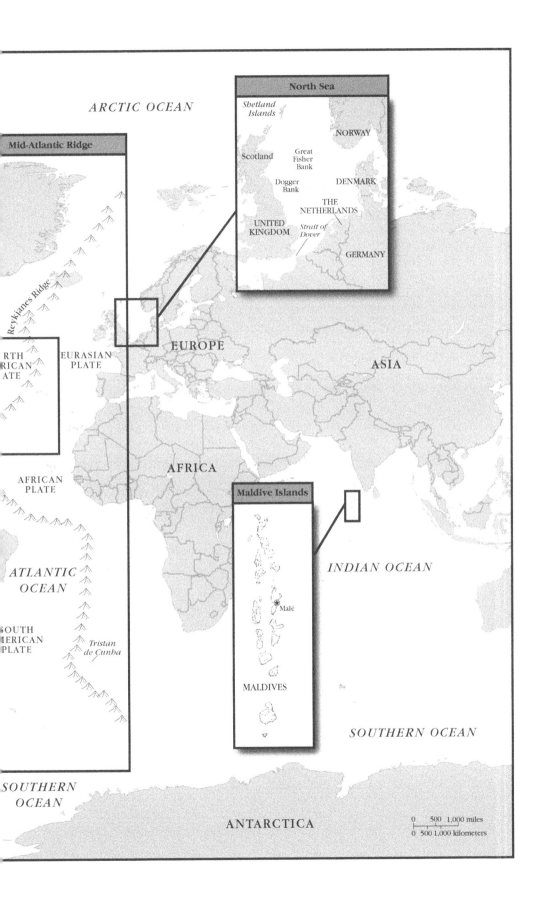

ARCTIC OCEAN

North Sea

Shetland Islands

NORWAY

Scotland

Great Fisher Bank

Dogger Bank

DENMARK

THE NETHERLANDS

UNITED KINGDOM

Strait of Dover

GERMANY

Mid-Atlantic Ridge

Reykjanes Ridge

EUROPE

ASIA

NORTH AMERICAN PLATE

EURASIAN PLATE

AFRICA

AFRICAN PLATE

Maldive Islands

INDIAN OCEAN

ATLANTIC OCEAN

SOUTH AMERICAN PLATE

Tristan de Cunha

Malé

SOUTHERN OCEAN

MALDIVES

SOUTHERN OCEAN

ANTARCTICA

0   500 1,000 miles
0 500 1,000 kilometers

4 Gulf of Alaska Pacific Ocean

Situated in the northeastern Pacific Ocean and known for its cold, deep waters, the Gulf of Alaska is a wide inlet tucked into the northwestern edge of North America. Spanning up to 592,000 square miles (1.5 million square kilometers), the gulf is bordered by land on two sides. To the east, north, and west is the state of Alaska, and to the south is the open Pacific Ocean. Where the gulf and land meet, the water depths are relatively shallow—between 300 and 600 feet (91 and 183 meters)—until a few hundred miles offshore, where the gulf drops sharply off the continental shelf to a maximum depth of around 15,000 feet (4,572 meters).

This region of the Pacific Ocean sits at the junction of warm and cold water currents, providing excellent opportunities for numerous species to thrive on the upwelling of nutrients. The gulf features a large circular current, or gyre, called the Alaska Current, which rotates in counterclockwise motion and carries waters from east to west. A second influence, the Kuroshio Current, contributes water from the southwestern Pacific, carrying warm, nutrient-rich water into the gulf. Phytoplankton living in the cold waters, nourished by the nutrients in the warm waters, grow and multiply to provide food for small fish, which form the basis for the regional marine food chain.

Considerable weather systems are formed in the gulf where warm waters from the Pacific Ocean intersect with cold waters from the north. This area has been described as the burial ground for western Pacific storms, which arrive low in energy after traveling across the ocean. When these storms and powerful surface currents in the northern Pacific come into contact with colder air, however, the condensation that forms in the atmosphere powers up new large storms that

These humpback whales are bubble net feeding in the Gulf of Alaska. In a complex, and not completely understood, social maneuver, a group of whales works together to encircle a school of fish above them by releasing a cylindrical wall of bubbles that functions like a net, temporarily trapping the fish. The whales then swim up through the cylinder, catching the fish in their huge mouths with a scooping motion. *(Ralph Lee Hopkins/ National Geographic/Getty Images)*

churn in the gulf and then rain and snow heavily on the adjacent coastal region. For example, in the nearby southern Alaskan city of Juneau, it rains up to 250 days per year, with an annual precipitation of up to 85 inches (216 centimeters). Storms that originate in the gulf impact much of the weather of northwestern North America, often carrying bands of rain across the land to Vancouver, British Columbia, in Canada and Seattle, Washington, in the United States.

The undersea geology in the Gulf of Alaska features a series of submerged dormant volcanoes called seamounts. Nearly 100 mountains, which were formed beginning 100 million years ago, span a 400-mile-long (644-kilometer-long) chain. Most of the peaks tower up to 12,000 feet (3,658 meters) from the bottom of the gulf, but all stop within 2,500 feet (762 meters) of the surface.

The seamounts support a diversity of large corals that provide distinctive habitat and shelter for crustacean and fish populations at various levels. In 2002 and 2004, crews in deep-sea submersibles, including the Woods Hole Oceanographic Institute's *Alvin*, turned their attention to these previously unstudied features.

EXAMPLES OF MARINE LIFE IN THE GULF OF ALASKA

Fish	Shellfish	Birds	Mammals	Reptiles
Pacific Cod (*Gadus macrocephalus*)	Abalone (*Haliotis discus hannai*)	Black-footed Albatross (*Phoebastria nigripes*)	Blue Whale (*Balaenoptera musculus*)	Green Sea Turtle (*Chelonia mydas*)
Herring (*Clupea pallasii*)	Red King Crab (*Paralithodes camtschaticus*)	Arctic Tern (*Sterna paradisaea*)	Harbor Seal (*Phoca vitulina*)	Leatherback Turtle (*Dermochelys coriacea*)
Mackerel (*Scomber japonicus*)	Pacific Sea Scallop (*Patinopecten caurinus*)	Parakeet Auklet (*Aethia psittacula*)	Killer Whale (*Orcinus orca*)	Olive Ridley Turtle (*Lepidochelys olivacea*)
Alaska Pollock (*Theragra chalcogramma*)	Shrimp (*Pandalus borealis*)	King Eider (*Somateria spectabilis*)	Sperm Whale (*Physeter macrocephalus*)	Loggerhead Sea Turtle (*Caretta caretta*)
Sardine (*Sardinops sagax*)	Snow Crab (*Chionoecetes opilio*)	Parasitic Jaeger (*Stercorarius parasiticus*)	Walrus (*Odobenus rosmarus*)	Hawksbill Turtle (*Eretmochelys imbricata*)

Sources: Alaska Natural Heritage Program and Alaska Department of Fish and Game, 2009.

Human Uses

Although no native people live in the middle of the landless Gulf of Alaska, various peoples have settled in the coastal regions as far back as 14,000 years ago. At that time, humans migrated from Asia across the Bering Land Bridge, which joined present-day Alaska and eastern Siberia at times during the Pleistocene ice ages.

Today, the Tlingit people, hunter-gatherers who number 11,000, live in south-eastern Alaska. Their diets rely on a steady catch of shellfish and gulf fish, primarily salmon. Their identity and culture are closely related to the marine environment.

Another ethnic group, the Athabaskans, occupies a region on the north end of the gulf. Seven separate subgroups live near the water. They sometimes use small fishing boats of the kind their ancestors developed centuries ago to catch bass, cod, mackerel, and salmon.

Due to the bountiful marine life in the gulf waters, the fishing industry has grown steadily since the 1970s. In 2000, the commercial industry employed

65,000 people who caught 4.4 billion pounds (2 billion kilograms) of fish and earned up to $2 billion.

The industry has focused on catching and selling a dozen popular species, including Pacific halibut (*Hippoglossus stenolepis*; 85 million pounds, or 39 million kilograms, per year), salmon (several species from the family *Salmonidae*; 170 million pounds, or 77 million kilograms, per year), and Alaskan pollock (*Theragra chalcogramma*; 1.7 billion pounds, or 800 million kilograms, per year).

Overfishing of red king crab (*Paralithodes camtschaticus*) in 1983 resulted in an 85 percent reduction in the species and strict fishing regulations. Despite catch limits, in the 2006 season, 250 boats caught 14 million pounds (6.4 million kilograms) of crab in less than a week, resulting in the closure of the season. Due to limited fishing opportunities, the fleet shrank to eighty-nine boats in 2008.

While fishing in the Gulf of Alaska can be lucrative, these turbulent waters are some of the most dangerous in the United States. According to the U.S. Bureau of Labor Statistics, the fatality rate is 151 deaths out of every 100,000 people employed in this industry. Drowning is the cause of up to 80 percent of the annual deaths.

Pollution and Damage

The Gulf of Alaska's dynamic geology presents a natural hazard to the region in the form of underwater earthquakes that cause undersea tidal waves, or tsunamis. The source is the northern portion of the North American tectonic plate colliding with the Pacific Plate.

On March 27, 1965, the Good Friday Earthquake, with a magnitude of 9.2, occurred; the epicenter was near Prince William Sound at the northern end of the gulf. This was the largest earthquake ever recorded in North America. It was equal to the force of 63,000 atomic bombs, and the sea floor was raised 36 feet (11 meters) in places. The event generated an enormous tsunami that was estimated at 1,742 feet (531 meters) high when it impacted southern Alaska at Lituya Bay.

The earthquake, which lasted for up to five minutes, killed 121 people on land, and the resulting waves killed another eleven, some as far away as Crescent City, California. The undersea fault that was formed measured 500 miles (805 kilometers) east to west by 125 miles (201 kilometers) north to south.

On March 24, 1989, at approximately midnight, the 987-foot (301-meter) oil tanker *Exxon Valdez* ran into a coastal reef in the Gulf of Alaska's Prince William Sound. The ensuing spill of 10.8 million gallons (41 million liters) of crude oil was one of the worst environmental disasters in U.S. history. The oil contaminated 1,500 miles (2,414 kilometers) of coastline and killed a huge number of marine animals, including an estimated half a million seabirds, 5,500 sea otters, 300 harbor seals, 300 bald eagles, and thirteen killer whales, as well as several billion individual fish and numerous other marine species.

The state of Alaska and the federal government reached a $900 million settlement with the Exxon Corporation in 1991. In 1994, Exxon paid $287 million to thousands of Alaska plaintiffs for actual damages to the environment and economy. Since the settlement, the oil company (now Exxon Mobil Corporation) has spent millions of dollars fighting an additional $5 billion penalty imposed for punitive damages. The U.S. Supreme Court rejected a lower court decision in 2008, sending the case back to an appeals court for a revised penalty amount.

In 2009, Exxon Mobil announced that it would not appeal the decision of the appeals court and that it planned to pay a judgment of $507 million and $470 million in interest. The corporation is refusing, however, to pay for the $70 million in legal fees claimed by the plaintiffs.

Following the *Exxon Valdez* oil spill of March 1989, one baby and five adult oil-soaked sea otters lay dead on the shores of Green Island, in Prince William Sound off Alaska. In 1990, the U.S. Congress enacted legislation requiring all oil tankers in U.S. waters to be double-hulled by the year 2015, in order to provide added protection against such environmental disasters. *(Chris Wilkins/AFP/Getty Images)*

The Gulf of Alaska's salt water has incurred harmful chemical alterations as a result of increased amounts of carbon dioxide and overall global climate change. Specifically, the lowered pH of the ocean water has affected the region's population of pteropods (*Limacina helicina*), often called sea butterflies, a species of very small mollusks that swims with wing-like flaps on the surface and eat plankton. Sometimes, the mollusks are unable to form their shells, which are made primarily of carbonate, due to the lack of calcium in the water and its high acidity.

Without the proper nutrients in the water, zooplankton bodies also are dissolving. The population decline of this key species, which feeds small fish, has begun to impact other species in the food chain.

Mitigation and Management

Beginning in the 1950s, the region's rapidly growing commercial fishing industry brought more and larger boats to scour the ocean waters in the gulf. With the populations of species such as coho salmon (*Oncorhynchus kisutch*) and Pacific Ocean perch (*Sebastes alutus*) in decline under these pressures, public outcry led to new laws to address overfishing in the Gulf of Alaska.

In 1976, the U.S. Congress passed the Fishery Conservation and Management Act to govern marine fisheries management and better conserve fish populations, also called stocks. This law was passed as multiple species in coastal waters, such as the sockeye salmon (*Oncorhynchus nerka*), began to disappear. Regional fishery management councils were established in the 1980s to draw up detailed management plans. The local affiliate, the Northern Pacific Fishery Management Council, drafted action plans for the gulf, including one to help restore the ground fish population, which includes up to thirty species.

This broad federal law was reauthorized in 2006 with a series of additional standards to enhance conservation efforts. In particular, the law sets a deadline of 2011 to stop overfishing by requiring catch limits, changing the way quotas are assigned, increasing enforcement efforts and federal and state partnerships in quota areas, establishing a new program to better study and protect deep-sea corals, and requiring fishery councils to establish scientific committees to provide data for future planning.

In an effort to better understand the diversity and biotic health of the Gulf of Alaska, the National Oceanographic and Atmospheric Administration (NOAA), a U.S. federal agency, declared the gulf region part of the Large Marine Ecosystems Program (LME) in 1984. This designation provided funding for gulf studies that use an integrated approach to sustaining the water body and its resources. Program studies examine ecosystems in five areas: socioeconomics, governance, pollution/ecosystem health, fish/fisheries, and ecosystem productivity.

Signs of climate change in the Gulf of Alaska have motivated new research to better understand this phenomenon. In the 1990s, the northern Pacific Ocean was included in the Global Ocean Ecosystem Dynamics (GLOBEC) program funded by the National Science Foundation and NOAA. The program supports research that studies how elements of the marine ecosystem respond to climate change—both today and in the past. Studies determine how changes in chemistry, currents, and the movements of storms reduce nutrients or change conditions for species in the gulf and thus affect population levels and catch amounts.

As of 2009, studies showed that the gulf's water temperature increase had led to a reduction in shrimp and crab and an increase in ground fish and salmon. When temperatures rise, nutrients are delivered to species that feed above the ocean bottom. This short-term data helps marine researchers understand possible long-term impacts of climate change on the ecosystem in the Gulf of Alaska.

Selected Web Sites

Alaska Department of Fish and Game: http://www.adfg.state.ak.us.
Alaska Marine Conservation Council: http://www.akmarine.org.
Alaska Oceans Program: http://www.alaskaoceans.net.
National Oceanic and Atmospheric Administration, Ocean Explorer, Alaska's Seamounts: http://oceanexplorer.noaa.gov/explorations/04alaska/.
North Pacific Fishery Management Council: http://www.fakr.noaa.gov/npfmc.
United Nations Large Marine Ecosystems of the World: http://www.edc.uri.edu/lme/text/gulf-of-alaska.htm.

Further Reading

Lieberman, Bruce. "Changing Ocean Chemistry Threatens to Harm Marine Life." *San Diego Union-Tribune,* September 14, 2006.
Mundy, Phillip, ed. *The Gulf of Alaska: Biology and Oceanography.* Fairbanks: Alaska Sea Grant College Program/University of Alaska at Fairbanks, 2005.
National Oceanographic and Atmospheric Administration. *Magnuson-Stevens Fishery Conservation and Management Reauthorization Act of 2006: An Overview.* Washington, DC: U.S. Department of Commerce, 2007.

National Research Council. *A Century of Ecosystem Science: Planning Long-Term Research in the Gulf of Alaska.* Washington, DC: National Academy Press, 2002.

North Pacific Fishery Management Council. *Central Gulf of Alaska Rockfish Demonstration Program.* Washington, DC: U.S. National Marine Fisheries Service, 2007.

O'Clair, Rita M., Robert H. Armstrong, and Richard Carstensen, eds. *The Nature of Southeast Alaska: A Guide to Plants, Animals and Habitats.* Portland, OR: Alaska Northwest, 2003.

Spies, Robert. *Long-Term Ecological Change in the Northern Gulf of Alaska.* Amsterdam, The Netherlands: Elsevier Science, 2007.

5 Mid-Atlantic Ridge
Atlantic Ocean

At the surface, the middle of the Atlantic Ocean is both an isolated and undistinguished area. But underneath thousands of feet of pitch-black salt water sits the Mid-Atlantic Ridge, a divergent tectonic plate boundary located along the ocean floor and the longest mountain range on Earth.

This 8,800-mile-long (14,159-kilometer-long) ridge, which has approximately 10 million individual peaks, runs in an S shape from the Arctic Circle to the Southern Ocean. In the North Atlantic, it separates the Eurasian Plate and the North American Plate, and in the South Atlantic, it divides the African Plate from the South American Plate. Although the Mid-Atlantic Ridge exists mostly deep underwater, portions of it rise between 1,000 and 12,000 feet (305 and 3,658 meters) below sea level.

The earthquake-prone ridge consists of two parallel sets of mountains, each up to 10 miles (16 kilometers) wide, transected by thousands of smaller transverse ridges. In the heart of the undersea terrain is a singular valley called the wound of the ridge, because this is where new ocean plates emerge. It is roughly 5 to 40 miles (8 to 64 kilometers) wide and hosts the Atlantic's deepest waters, exceeding 25,400 feet (7,742 meters).

This geologic formation is the product of magma that is emerging from within a fracture of the tectonic plate. Known as "sea floor spreading," the new ocean floor has resulted in the growth of the Atlantic Ocean. The ridge is growing up to 2 inches (5 centimeters) in size per year, both toward the west and the east.

The geology of the Mid-Atlantic Ridge dates to the Permian Period, when dinosaurs roamed the Earth. All of the continents were grouped into one C-shaped mass termed Pangaea, surrounded by the single ocean, Panthalassa.

Approximately 225 million years ago, the movement of the tectonic plates caused this single continent to begin breaking into seven pieces. By 200 million years ago, a 2,500-mile-long (4,023-kilometer-long) rift had formed between South America and Africa. What began as a small saltwater estuary, only a half mile (0.8 kilometer) wide, eventually grew into a divergent oceanic boundary, as magma emerged from under the bedrock to form new sea floor.

The middle of the Atlantic Ocean contains dozens of currents. These currents move in multiple directions at any one time, based on the time of year, location, depth, winds, and water temperature. In addition, the height and size of the Mid-Atlantic Ridge alter the velocity and direction of these currents, either by redirecting, deflecting, or channeling water movement. Three prominent larger currents include

1) A northern current, where subarctic cold waters swirl in a clockwise direction (from North America northeast toward Europe)

2) An equatorial current in the middle of the ocean, where warmer waters travel from east to west (from Africa to South America)

3) A southern current at the edge of the Atlantic and Southern oceans, where colder waters swirl in a counterclockwise circle (from South America to Africa)

The rocks of the Mid-Atlantic Ridge contain historical records embedded deep within their grains. One is the periodic shift of the magnetic polarity of the Earth. The planet contains a large amount of metal, especially in its core, and this metal puts off an electromagnetic field that radiates into deep space. There are times when the polarity of this magnetic field has "flipped" from showing magnetic north (which is the current polarity) to magnetic south. By dragging a magnetometer behind an ocean ship, researchers in the 1960s found long narrow bands of rock that reveal such a periodic change in polarity. This shift has occurred at least 176 times in the last 200 million years, or about once every 500,000 years.

Scientists do not completely understand this phenomenon. One theory is that these geomagnetic reversals may be triggered when the Earth's magnetic field becomes weaker, allowing a shift in poles.

Human Uses

Up until the 1940s, little was known about the Mid-Atlantic Ridge. At that time, early versions of electronic sonar (sound navigation and ranging), a technology that uses sound waves to measure solid objects underwater, were used to measure ocean depths and basic underwater land formations. This was a slow process, given the enormity of the ocean.

In 1947, Marie Tharp and Bruce Heezen, two oceanographers from Columbia University in New York City, began working on a project to map the world's oceans. Heezen gathered the data while Tharp put the numbers into map form. By 1961, they had completed maps of both the North and South Atlantic and determined that a substantial ridge with associated valleys ran the length of the ocean.

By the 1970s, space-based satellites allowed for remote sensing to produce high-resolution images and other scientific information with significantly less time and cost than earlier methods. In addition to mapping of ocean topography, this technology also gathered such data as water currents, temperature, and wave height.

Despite its size, the Mid-Atlantic Ridge features only a few groups of islands along its length. An archipelago of six islands sits in the southern Atlantic Ocean; they comprise some of the most remote locations in the world. The largest island, Tristan da Cunha, is 38 square miles (98 square kilometers) in size. The closest continents to this British-owned but independently administered territory are South America (2,080 miles, or 3,347 kilometers, away) and Africa (1,750 miles, or 2,816 kilometers, away). The island's tallest point is a 6,765-foot (2,062-meter) dormant volcano, which last erupted in 1960.

Historically used as a stopping point for sailing ships, Tristan da Cunha supports a population of fewer than 300 people. The island has no airport and can only be reached by boat; supplies often are delivered by South African fishing vessels.

Residents rely on the ocean for survival. Day boats pull up traps for Tristan rock lobsters (*Jasus tristani*), or saltwater crayfish, which are frozen or canned and exported as a source of income. After overfishing caused a significant decline in the island's lobster population in the 1990s, the British government implemented a strict quota system—a limit of 395 tons per year in 2008—to sustain the fishery.

The island of Iceland is a landmass formed by the Mid-Atlantic Ridge. Located at the northern edge of the Atlantic Ocean between Greenland and northern Europe, this 39,770-square-mile (103,004-square-kilometer) island was inhabited by 317,593 residents as of 2009.

Iceland sits at the edge of the Arctic Circle but is bathed periodically by the warmer Gulf Current, making its weather milder. The island has at least a dozen active volcanoes. Eruptions on Mount Laki in 1783 took the lives of approximately 12,000 residents. Hekla, a 4,882-foot (1,488-meter) volcano in southern Iceland, has erupted more than twenty times since 874 C.E. The three most recent eruptions, in 1980, 1991, and 2000, covered portions of Iceland in ash and melted much of the surrounding snow but produced little lava flow.

In April 2010, the volcanic eruption of the 5,466-foot (1,666-meter) volcano Eyjafjallajökull (its name means "glacier of the island-mountains") on Asolfsskali, Iceland, resulted in a plume of volcanic ash that reached elevations of 4.9 miles (8 kilometers) in altitude. This caused such widespread air pollution across Europe that almost all airline flights were halted for five days, resulting in estimated economic losses of $1.5 billion. *(Emmanuel Dunand/AFB/Getty Images)*

The numerous volcanoes provide a unique environment for the island's five geothermal plants, which utilize the heat of the Earth's interior to produce 26 percent of the country's electricity, and 74 percent of the island's power comes from hydroelectric facilities powered by glacial runoff. Fossil fuels, such as oil, are only used for backup generation during periods of reduced runoff.

Iceland also benefits greatly from the intersection of cold water (from the Arctic Ocean) and warm water (from the Caribbean Sea), which supports diverse marine life, including 320 fish species. In 2005, the fishing industry on the island was the twelfth largest in the world, with more than 1.6 million tons of fish landed annually, including herring (*Clupea harengus*) and Atlantic cod (*Gadus morhua*).

In order to better manage fish stocks, the Icelandic government began systematically closing areas for a period of time in 1975. It also increased its oversight with the 1990 Fisheries Management Act, which promotes the conservation and efficient utilization of marine stocks. Following their ancestral traditions, Icelanders still engage in whaling, despite the International Whaling Commission's 1986 ban. In 2007, Icelanders hunted and killed thirty-nine common minke whales (*Balaenoptera acutorostrata*), mostly for their meat.

Pollution and Damage

Due to the Mid-Atlantic Ridge's remote location, human impacts, such as pollution from land runoff and direct garbage dumping, are not common. Passing ships that dump waste are direct, but minor, polluters. However, air pollution causes some impact to the Mid-Atlantic Ridge region. Carried aloft from several continents, carbon, sulfur, and other debris land in ocean waters and change water chemistry, thus impacting marine life. Although the middle of the Atlantic Ocean is a vast area, scientific evidence has documented acidification, which harms species such as clams, crabs, and shrimp that rely on the water's pH to grow their shells.

Recent volcanic activity in Iceland also has caused noteworthy pollution. During the first three days of its April 2010 eruption, Eyjafjallajökull spewed some 2.8 billion cubic feet (80 million cubic meters) of lava and 3.5 billion cubic feet (100 million cubic meters) of fine-grain ash. An eruption of this magnitude can have short-term impacts on global weather patterns, including reducing air temperatures as the ash cloud particulates interfere with solar heating of the planet.

The volcanic processes along the ridge also regularly alter the deep underwater environment. When 35-degree Fahrenheit (1.6-degree Celsius) seawater is drawn into volcanic fissures, hot magma and gasses eject it with significant pressure. The superheated water (up to 750 degrees Fahrenheit, or 399 degrees Celsius) flows upward in large plumes full of debris, termed "black smokers" for their smokestack-like appearance. These hydrothermal vents, only discovered by research submersibles in 1979, support a diversity of sea life in what scientists once thought were uninhabitable spots at the bottom of the oceans.

In fact, the vented water is full of dissolved minerals such as iron and sulfur that feed adjacent bacteria in a process called chemosynthesis. These colonies grow rapidly into a thick layer and then are consumed by creatures such as amphipods (*Calyptogena sp.*), which, in turn, provide food for crabs, tube worms, and snails. One noteworthy ridge species is the deep-sea octopus (*Bathypolypus arcticus*), an eight-armed cephalopod known for its intelligence and skills in catching a wide range of ocean-bottom prey.

Mitigation and Management

Deep-sea dives conducted since the 1970s have revealed a distinct environment along the Mid-Atlantic Ridge. In 1974, France and the United States launched the French-American Mid-Ocean Undersea Study, a joint project that managed some sixty deep-sea dives in a section of the ridge about 200 miles (322 kilometers) southwest of the Azore Islands, which are 900 miles (1,448 kilometers) west of Portugal.

The *Alvin* submersible research station from Woods Hole Oceanographic Institution and the *Archimede* and *Cyana* submersibles from the French Research Institute for Exploration of the Sea dove together. Plunging to nearly 11,000 feet (3,353 meters), a depth at which the pressure equals 2 tons per square inch, submersible pilots were surprised when powerful currents pushed their reinforced hulls sharply up or down slopes.

The project sampled the chemical composition of the superhot water emerging from hydrothermal vents. Analysis revealed a mixture of dissolved metals such as chromite, copper, and manganese, a combination lethal to humans. More than 150 new species were documented in these historic dives.

In 1997, a twenty-three-year reunion took place. Meeting again near the Azore Islands, the French *Nautile* and the American *Alvin* explored two-dozen hydrothermal vents along the ridge. This time, scientists chronicled the existence of approximately 300 previously unknown species, including clams and mussels measuring 1 foot long (0.3 meters long) and tube worms that grow up to 10 feet (3 meters) in length.

In an effort to better inventory the marine life living along the Mid-Atlantic Ridge, in 2004 sixteen nations agreed to participate in a study of the region's ecosystems. Named MAR-ECO (Mid-Atlantic Ridge and ECO, referring to ecology), the project's central goal was to examine the distribution, abundance, and trophic (how one species gets nutrition from another) relationships of the

Mid-Atlantic marine species. Overseen by the Institute of Marine Research and the University of Bergen in Norway, vessels and scientists worked collaboratively to gather information about the northern Atlantic Ocean between Iceland and the Azores. Crews observed and studied various marine species and birds, including hundreds of dolphins, sperm whales, sei whales, finback whales, fulmars, and greater shearwaters.

Despite progress in learning more about the undersea biomes, the deep sea region along the Mid-Atlantic Ridge also is an extremely dangerous environment. In March 2010, the fifteen-year-old Autonomous Benthic Explorer (ABE), one of the most well-known untethered submersible vehicles to explore the ridge, was lost off the coast of Chile. It was theorized that, during the submersible's 222nd dive, strong currents may have pushed it into a rock surface, causing a buoyancy sphere to implode. Such an implosion would have exerted extreme pressure of more than 2 tons per square inch on the craft, resulting in catastrophic failure at depths greater than 8,842 feet (2,695 meters).

As the Mid-Atlantic Ridge has been explored, so have other chains of underwater mountains in the Pacific, Southern, and Indian oceans. Scientists have theorized that the Mid-Atlantic Ridge is part of a continuous singular undersea ridge that spans 40,000 miles (64,360 kilometers) around the planet. At the southern terminus of the Mid-Atlantic Ridge, the ridge becomes the Southwest Indian Ridge in the Indian Ocean. This ridge links to the Central Indian Ridge and continues farther east into the Pacific Ocean, connecting to the Pacific-Antarctic Ridge and, eventually, farther north, to the East Pacific Rise off western North America.

These ridges are geologically active and are always changing. As of the early twenty-first century, they continue to expand the sea floor from deep magma chambers at the bottom of oceans around the world.

Selected Web Sites

Geology of Iceland: http://www.iceland.is/country-and-nature/nature/Geology/.
Mid-Atlantic Ridge Ecology Study (MAR-ECO): http://www.mar-eco.no.
U.S. Geological Survey, Ocean Floor Mapping: http://pubs.usgs.gov/gip/dynamic/developing.html#anchor10564457.
U.S. Geological Survey, *This Dynamic Earth: The Story of Plate Tectonics:* http://pubs.usgs.gov/gip/dynamic/dynamic.html.
University of Guelph, Canada's Aquatic Environments, Mid-Atlantic Ridge: http://www.aquatic.uoguelph.ca/oceans/AtlanticOceanWeb/AOFloor/MARidge.htm.

Further Reading

Crist, Darlene Trew, Gail Scowcroft, and James M. Harding, Jr. *World Ocean Census: A Global Survey of Marine Life*. Buffalo, NY: Firefly, 2009.

Hutchinson, Stephen. *Oceans: A Visual Guide*. Buffalo, NY: Firefly, 2005.

Leier, Manfred. *World Atlas of the Oceans*. Buffalo, NY: Firefly, 2001.

Nordic Working Group on Fisheries Research. *The Mid-Atlantic Ridge Is Teeming With Life*. Copenhagen, Denmark: Nordic Council of Ministers, March 2005.

Pauly, Daniel, and Jay Maclean. *In a Perfect Ocean: The State of Fisheries and Ecosystems in the North Atlantic Ocean*. Washington, DC: Island, 2003.

Robinson, Ian. *Measuring the Oceans From Space*. New York: Springer, 2004.

Sverdrup, Keith. *Introduction to the World's Oceans*. New York: McGraw-Hill Higher Education, 2008.

Tove, Michael H. *Guide to the Offshore Wildlife of the Northern Atlantic*. Austin: University of Texas Press, 2001.

6 Maldive Islands Indian Ocean

The Maldive Islands are located 430 miles (692 kilometers) southwest of Sri Lanka in the central Indian Ocean. The islands are spread out in a north-south-oriented rectangular patch across 34,749 square miles (90,000 square kilometers) of remote ocean territory—one of the most spread out countries on the planet. The Maldives are made up of 1,192 small islets, or atolls. Eight hundred of them have some vegetation, while only 200 are large enough to support human habitation.

Spanning 620 miles (998 kilometers) from north to south, the islands sit atop the mostly underwater Chagos-Laccadive submarine ridge. The ridge, made up of more than 100 prominent undersea volcanoes, runs for 1,850 miles (2,977 kilometers) from western India into the deep central Indian Ocean. The ocean depths adjacent to the Maldive Islands are less than 3,200 feet (975 meters); however, at the southern end of the ridge, the Indian Ocean drops to 16,300 feet (4,968 meters) deep.

Daytime temperatures on these tropical islands average 82 degrees Fahrenheit (28 degrees Celsius), and it rains 78 inches (198 centimeters) per year. This climate supports a tropical rain forest in the upland locations. The maximum elevation is 16 feet (5 meters) above sea level, and the average elevation is less than 5 feet (1.5 meters), which poses major risks for inhabitants, such as coastal flooding and limited freshwater supplies.

The islands underwent three stages of development, the first of which began during the Miocene Epoch, roughly 25 million years ago. Volcanoes rose up from the ocean floor to form a ridge and, occasionally, a circular island above the water. Second, these landforms collapsed after 5 to 10 million years of erosion, leaving an outer ring of submerged rock and a central shallow lagoon. Most atolls measure about 0.5 mile, or 0.8 kilometer, in diameter and occur in clusters, with 60 percent in the Pacific Ocean, 39 percent in the Indian Ocean, and less than 1 percent in the Atlantic Ocean.

The third stage of island formation occurred over hundreds of years, as dozens of species of coral—a living animal that secretes calcium carbonate to form a hard exterior skeleton—established colonies on the ring of rock, both above and below the ocean surface. Often growing up to 8 feet (2.4 meters) high and 30 feet (9 meters) wide, the coral provides habitat for marine life and a level of protection from wave action for the island. Even after coral dies, its skeleton remains, and new coral then grows on top of the original mass, creating an impressively large structure over time.

The Maldives coral reefs are the seventh largest on Earth. They span 5,540 square miles (14,349 square kilometers) and consist of 187 species of coral. These corals live in relatively shallow, low-nutrient, warm water (above 68 degrees Fahrenheit, or 20 degrees Celsius), feed on plankton and algae, and grow up to 0.25 inch (about 0.64 centimeters) in diameter per year. In areas of intense wave action, corals may be dense and smaller in order to withstand the forces of the ocean. Away from the surf, in somewhat deeper waters, corals grow in elongated, delicate shapes and spread-out formations. Scientists have described them as an "underwater rain forest" that provides an ideal habitat for approximately 1,090 tropical species of crustaceans, fish, and mollusks.

CORAL REEF FACTS

- Percentage of marine life that calls a coral reef home: 25 percent
- Estimated species of coral, planetwide: 1,000
- Amount of reefs lost to development or impacts from climate change since 1970, in square miles: 54,687 (141,639 square kilometers)
- Estimate of reef mass that will be lost in next eighty years: 75 percent

Source: Coral Reef Conservation Program, U.S. National Oceanic and Atmospheric Administration, 2008.

EXAMPLES OF WILDLIFE IN THE MALDIVES

Mammals (28 Species Identified)	Fish (1,090 Species Identified)	Reptiles (12 Species Identified)	Birds (170 Species Identified)
Blue Whale (*Balaenoptera musculus*)	Butterfly Fish (*Chaetodon mitratus*)	Blind Snake (*Ramphotyphlops braminus*)	Common Cuckoo (*Cuculus canorus*)
Fruit Bat (*Pteropus giganteus ariel*)	Green Parrot Fish (*Leptoscarus vaigiensis*)	Green Sea Turtle (*Chelonia mydas*)	Pacific Golden Plover (*Pluvialis fulva*)
Small Flying Fox (*Pteropus hypomelanus*)	Grey Reef Shark (*Carcharhinus amblyrhynchos*)	Hawksbill Turtle (*Eretmochelys imbricata*)	Roseate Tern (*Sterna dougallii*)
Sperm Whale (*Physeter macrocephalus*)	Manta Ray (*Manta birostris*)	Indo-Pacific Gecko (*Hemidactylus garnotii*)	Ruddy Turnstone (*Arenaria interpres*)
Striped Dolphin (*Stenella coeruleoalba*)	Rock Cod (*Cephalopholis sexmaculata*)	Olive Ridley Turtle (*Lepidochelys olivacea*)	Short-Eared Owl (*Asio flammeus*)

Source: Republic of Maldives, 2008.

The most common types of coral in the Maldives are stony corals (from the order *Scleractinia*), which are known for their bright colors. These tiny, individual animals consume microscopic algae and zooplankton, and their bodies secrete a calcium-based mineral that eventually forms a hollow but hard calcium-based structure. Another type of coral—called brain coral (from the family *Faviidae*), due to their grooved surface, which resembles an animal brain—can live up to 900 years. Most brain coral colonies are about 5 feet (1.5 meters) high by 4 feet (1.2 meters) wide and consist of thousands of living individuals.

Marine species that establish themselves in coral colonies include cnidarians (jellyfish), crustaceans (crabs), echinoderms (sea urchins and starfish), fish (angelfish, grouper, manta, parrot fish, sharks, and snapper), mammals (dolphins), mollusks (clams, conch, and snails), and reptiles (sea snakes and sea turtles).

Human Uses

The Maldive Islands have been occupied by humans since the fifth century, when small groups of Indian Hindus traveled via boat to settle the area. In 1153, Muslim Arabians arrived to spread Islam (it still is the majority religion on the islands). They were followed by the Portuguese, who ruled beginning in 1558, and then by the British, who ruled from 1887. The modern-day residents' ethnic heritage includes a mix of African, Arabic, European, Indian, and South Asian cultures. The common language, Dhivehi, has roots in several tongues.

In 1953, the Maldives became an independent republic. President Maumoon Abdul Gayoom held office from 1978 to 2008, despite several attempted overthrows. He was defeated in the October 2008 presidential election by former member of parliament Malé Mohamed Nasheed of the Maldivian Democratic Party.

In 2009, 396,334 residents lived on the islands. Up until the 1930s the population was approximately 100,000, but as the local tourism economy grew so did the number of residents. The population tripled in seventy years' time, though the growth rate slowed to 1.8 percent in 2009. Although the majority of the islands are not occupied by humans, seventy of the habitable islands have fewer than 500 residents.

Located at the southern edge of North Malé Atoll, or Kaafu Atoll, Malé, the capital and largest city in the Republic of Maldives, has 103,693 residents that occupy an area that is 1 mile by 0.62 miles (1.6 kilometers by 0.99 kilometers) wide. Malé is covered with buildings, and the entire island is surrounded by a 10-foot-high (3-meter-high) seawall that cost $63 million to construct in 1987; 99 percent of it was paid for by Japan, after widespread flooding occurred and the Maldives government could not afford the wall to protect its residents. Several landfill projects over the years have expanded the island and harbor, creating new roadways and protected waterways.

All of the islands have poor soil, lacking in manganese, nitrogen, and potash, with an average depth of less than 8 inches (20 centimeters). The nutrient-deficient soil makes farming prohibitive in most places; only 10 percent of the land supports crops, such as baraboa (a type of squash), bileiy (a leafy plant), fruit trees (bananas, coconuts, and papaya), and githeyo mirus (a type of hot chili pepper).

Historically, the economy of the Maldives has relied on fishing. Boats still are built from the only harvestable tree, the coconut palm (*Cocos nucifera*). The Malé fleet consists of 352 trawlers (also called dhoni) that hold twelve-person crews that often fish for tuna.

Since the 1980s, the economy has shifted and has become reliant on tourism. In 2004, 563,593 people visited the Maldives. Although this economic change has increased per capita income, it also has linked the island to the current events and economy of the broader international community. After the terrorist attacks of September 11, 2001, in the United States, for example, the hotel industry lost almost 90 percent of its business for several months, because travelers were not willing to fly due to safety concerns. Tourism also declined steeply after the

In this picture of an island in the Maldives, the coral reef visible in the background supports a diversity of species and also protects the island from wave impacts. Unfortunately, events such as a rise in water temperature and increasingly severe weather conditions over the past ten years have harmed the Maldives coral, resulting in the loss of important reefs and an increase in flooding and erosion. (© traveller/Fotolia)

December 2004 Indian Ocean tsunami, which caused massive flooding, coral and beach impacts, and the closure of the majority of the archipelago's resorts.

Pollution and Damage

Residents of the Maldives mine for submerged rock and coral to use for upland fill and building materials. As the population has grown, more rock and coral have been needed to construct office and housing complexes, which are built at 16-foot (4.9-meter) elevations to withstand the effects of ocean tides. Mining coral and rock, however, is unsustainable, because the harvest rate is too rapid, and excessive removal undermines existing atolls and reduces critical coral habitat. Between 1975 and 1985, 3.3 million cubic feet (93,000 cubic meters) of coral rock was removed per year; that figure had increased to 35 million cubic feet (1 million cubic meters) per year by 2009, primarily due to the construction of large resorts.

Natural disasters are a serious threat to the safety of Maldives residents. The earthquake that began underwater off the west coast of Sumatra (measuring 9 on the Richter Scale) at 3 P.M. on December 26, 2004, resulted in a massive tsunami that sped across the Indian Ocean at speeds of up to 500 miles per hour

(805 kilometers per hour) and landed on the shores of the Maldive Islands at 6:30 P.M. without sufficient warning. The resulting series of waves killed eighty-two residents, injured 1,230, and swept thirty-one out to deeper waters, where most were never found. The waves severely damaged 3,854 buildings, left 12,162 people homeless, damaged 120 boats, and swept another twenty-five boats out to sea. The World Health Organization's post-event assessment found that up to 31 percent of the island's structures were lost. In the aftermath, officials stated that Malé's seawall saved thousands of lives by deflecting the waves.

In 2006, a network of twenty-five seismographic stations and three deep-ocean sensors was activated in the Indian Ocean to serve as a future tsunami warning system. In 2009, the government had the region resurveyed, because the Maldive Islands were so altered by the impacts.

The island communities have little defense against the region's surging waters. In the last fifty years, several communities have reported wells and soil contaminated with salt water, widespread flooding, and severe coastal erosion, leaving little dry land.

In 1987, President Gayoom addressed the United Nations regarding such issues, calling his country "an endangered nation" and offering strong support of international efforts to address climate change. After being partially submerged after the 2004 tsunami, and facing the realization that the island may be completely submerged by the year 2040 (per the International Panel for Climate Change in 2007), 60 percent of the residents (2,100 people) on the island of Kandolhudhoo agreed to leave their homes by the year 2020.

VISION OF THE NATIONAL BIODIVERSITY STRATEGY AND ACTION PLAN OF THE MALDIVES

A nation which appreciates the true value of the natural environment, utilizes its natural resources in a sustainable manner for national development, conserves its biological diversity, shares equitably the benefits from its biological resources, has built the capacity to learn about its natural environment and leaves a healthy natural environment for future generations.

Source: Ministry of Home Affairs, Housing, and Environment, Republic of Maldives, 2002.

Climate change is a pressing threat to the Maldives. Warmer temperatures across the globe melt ice at the poles and increase ocean levels. Although the highest upland area in the Maldives is 16 feet (4.8 meters) above sea level, three-quarters of the landmass is only 3 feet (0.9 meters) above high tide levels.

The U.S. National Aeronautics and Space Administration (NASA) reported that ocean surface temperatures around the world have warmed by over 1 degree Fahrenheit (0.6 degrees Celsius) on average since the late 1970s. The joint U.S. and French weather satellite *Poseidon* recorded an Indian Ocean water level increase of almost 0.5 inch (1.3 centimeters) since 2000 and, in the period between 1993 and 2003, an increase of 1.2 inches (3 centimeters) over previous years. If present rates continue, scientists estimate that sea levels will rise more than 1 foot (0.3 meters) during the twenty-first century.

Global climate change also has affected the area's coral. In 1998, many coral reefs in the Maldives and around the world suddenly changed from a range of colors, including reds, yellows, greens, and tans, to bright white, a condition termed bleaching that occurs when the algae that feed on the coral are killed due to a sudden change in ocean temperature (between 2 and 5 degrees Fahrenheit, or -16.7 and -15 degrees Celsius). The 1998 warming lasted for months, and up to 70 percent of the coral in the Maldives was lost. Despite this loss of hundreds of years of growth, many of the reefs were growing once again by 2005.

Mitigation and Management

Given limited resources of land and freshwater, the Maldives government is cognizant of the need to use sustainable policies to assure the long-term health of its island communities. Two laws have been implemented to address short- and long-term environmental goals. The 1987 Fisheries Act of Maldives enables government officials to set specific sustainable catch limits on species such as tuna, and the 1993 Environment Protection and Preservation Act sets standards to reduce land pollution and protect reef communities.

In addition, in 1992 the Republic of Maldives was the first nation to sign the United Nations Convention on Biological Diversity, an international treaty whose goals include the conservation of biological diversity and sustainable use of resources. In 2002, the republic completed the National Biodiversity Strategy and Action Plan of the Maldives. The dozen major objectives of the plan include integrating biodiversity conservation into any development on the island, teaching residents about the importance of resource and animal conservation, and em-powering residents through increased participation in community projects.

In 2005, in response to a request from President Gayoom, Australian Prime Minister John Howard committed a team of Australian marine scientists to assist scientists from the Maldives Marine Research Center in the assessment of damage to coral reefs following the 2004 tsunami. After seventeen days of surveys across 124 reef sites on seven atolls, the nine-member scientific team reached two primary conclusions. First, the waves from the tsunami had made a significant impact with considerable force, depositing millions of tons of sediment and damaging coral. Second, the coral reefs and baitfish, although harmed, were recovering from what appeared to be only short-term damage. The conclusions were released in a joint report prepared by the Australian Government Mission and the Maldives Marine Research Center in 2005.

Additional questions remain, however, regarding how the reefs will fare from future impacts of climate change in this delicate ocean community. In March 2009, to do their part toward combating global climate change, President Nasheed pledged to make the Maldive Islands carbon-neutral within a decade by converting his nation to wind and solar power.

Selected Web Sites

An Assessment of Damage to Maldivian Coral Reefs and Baitfish Populations from the Indian Ocean Tsunami: http://www.ausaid.gov.au/publications/pdf/maldives_reef_report.pdf.

Convention on Biological Diversity, Maldives: http://www.cbd.int/countries/default.shtml?country=mv.

U.S. National Oceanic and Atmospheric Administration Coral Reef Conservation: http://coralreef.noaa.gov.

United Nations Coral Bleaching: http://earthwatch.unep.net/emergingissues/oceans/coralbleaching.php.

United Nations Coral Reef Assessment: www.fao.org/docrep/x5627e/x5627e0a.htm.

Further Reading

Australian Government Mission and Maldives Marine Research Center. *Assessment of Damage to Maldivian Coral Reefs and Baitfish Populations from the Indian Ocean Tsunami*. Canberra, Australia: Commonwealth of Australia, 2005.

Bindoff, Nathaniel, et al. *Observations: Oceanic Climate Change and Sea Level*. New York: Cambridge University Press, 2007.

Dobbs, David. *Reef Madness: Charles Darwin, Alexander Agassiz, and the Meaning of Coral*. New York: Pantheon, 2005.

Fabricius, Katharina. *Marine Life of the Maldives*. Gig Harbor, WA: Sea Challengers, 2001.

Miller, Frederic, Agnes Vandome, and John McBrewster, eds. *Maldives*. Beau Bassin, Mauritius: Alphascript, 2009.

Antarctic Peninsula
Southern Ocean

The mixing of very deep warm and cold waters marks the intersection of the Atlantic, Pacific, and Southern Oceans at the southern terminus of the planet. This creates a biologically productive and simultaneously harsh environment. The Antarctic Peninsula juts 1,200 miles (1,931 kilometers) into the heart of this ocean junction; it is bordered on the west by the Bellingshausen Sea, on the north by the Drake Passage, and on the east by the Weddell Sea. The peninsula is 200 miles (322 kilometers) wide where it connects with the Antarctic mainland but ranges between 20 and 100 miles (32 and 161 kilometers) wide as it stretches northward into the Southern Ocean.

The northern tip of the Antarctic Peninsula sits 600 miles (965 kilometers) from South America. Due to its location 900 miles (1,448 kilometers) from the South Pole, however, the peninsula is covered by snow and ice except during the summer when coastal rocks are exposed. Three countries claim ownership of this landmass: Argentina, Chile, and Great Britain.

The formation of the peninsula dates back 165 million years to the Jurassic Period. A single supercontinent called Pangea existed, surrounded by Panthalassa, a single superocean. When this massive continent began to break up, the eastern portion—made up of Antarctica, Australia, India, and Madagascar—pushed away from Africa but stayed connected to the southern tip of South America. One by one, the other continental pieces broke off and moved away.

Around 80 million years ago, Antarctica still was connected to South America. At that time, this large continent began to rotate in a clockwise direction, and a large section of the eastern part of the landmass began to shear away, forming

HIGHEST PEAKS OF THE ANTARCTIC PENINSULA

Name	Location	Height (Feet)
Mount Jackson	Palmer Land	10,446
Welch Mountains	Palmer Land	9,892
Mount Stephenson	Alexander Island	9,800
Mount Egbert	Alexander Island	9,498
Mount Hope	Palmer Land	9,383

Source: Landsat Image Mosaic of Antarctica, U.S. National Aeronautics and Space Administration, 2009.

a long, narrow section of land. This motion, which ended 40 million years ago, formed the arm of land that is today termed the Antarctic Peninsula. As the Atlantic Ocean grows in size, Antarctica and South America are being pushed away from one another. The peninsula comprises 15 percent of the total continent of Antarctica.

The peninsula contains the second-longest mountain range on the continent, which runs its entire length. These peaks are geologically part of the Andes Mountains in South America. Where the Andes seemingly end in Chile, submarine mountains called the Scotia Ridge continue underwater for 2,700 miles (4,344 kilometers), forming a large curve to connect to the peninsula. Along the way, the ridge surfaces periodically, forming several island chains, including the South Sandwich Islands, the South Orkney Islands, and the South Shetland Islands. The peninsula hosts hundreds of mountains, the highest being Mount Andrew Jackson (10,446 feet, or 3,184 kilometers), which is located in Palmer Land, on the broad southern half of the peninsula.

The geology of this part of the planet has created a unique intersection of the Atlantic, Southern, and Pacific oceans, where varying currents and weather systems cross one another and support considerable biotic richness. The marine organisms that live in this tumultuous zone actually thrive where warm and cold waters intersect.

Powerful currents and storms result in nutrients rising from the ocean depths to the surface. Multiple phytoplankton species, relying on photosynthesis, consume these nutrients and thrive.

One species of phytoplankton, the unicellular diatom (*Chaetoceros dichaeta*), is the preferred food for krill (*Euphausia superba*), one of the larger species of zooplankton. These shrimp-like marine invertebrates consume algae so fast that their population swells to an astounding 500 million tons (more than two times the weight of all of the humans on the planet) in a single year. Baleen whales, such as the blue whale

Due to the amount of salt water that it ingests feeding on fish while it dives underwater, the Antarctic petrel (*Thalassoica Antarctica*) has a salt gland above its nasal passage that assists in desalinating its body. The gland releases the resulting high-salt solution out of the bird's nose. *(Frans Lemmens/The Image Bank/Getty Images)*

(*Balaenoptera musculus*), the largest known mammal, grow up to 110 feet (34 meters) long and 181 tons in weight and feed exclusively on krill.

The Southern Ocean also contains millions of pelagic birds, named from the Greek word *pelagikos,* meaning "sea." Over years of evolution, they have adapted to an open-water environment, and most species spend the greater part of their lives on and above the water. They fly to land only to mate, lay eggs, and briefly care for their young.

Such seabirds have remarkable feeding skills. Some dive up to 250 feet (76 meters) underwater to catch fish. One, the sooty shearwater (*Puffinus griseus*), feeds on fish and squid.

The Antarctic giant petrel (*Macronectes giganteus*) is the largest bird on the peninsula, with a length of 34 inches (86 centimeters) and a wingspan of 77 inches (196 centimeters). It feeds on the carcasses of fish, seals, penguins, or whales. This species of petrel does not mate until it is ten years old, and many live to be fifty, making it one of the longest-living ocean birds. However, the population of giant

PELAGIC BIRDS IN ANTARCTICA

Bird Name	Noteworthy Characteristics	Estimated Area Population
Australasian Gannet (*Morus serrator*)	While diving underwater for food, these birds inflate air sacs to protect their heads upon impact.	110,000
Snow Petrel (*Pagodroma nivea*)	Lives near drifting pack ice and icebergs; in flight, dips underwater to feed.	2,000
Sooty Shearwater (*Puffinus griseus*)	Feeds on a diet of small fish and squid; these birds can dive underwater up to 223 feet (68 meters) for food.	200,000
Southern Royal Albatross (*Diomedea epomophora*)	Second-largest of the species, able to migrate 9,300 miles (14,964 kilometers) without landing.	30,000
Wandering Albatross (*Diomedea exulans*)	With an 11-foot (3.35-meter) wingspan, glides for hours at a time with locked wings.	42,000

Source: Shirihai, Hadoram. *The Complete Guide to Antarctic Wildlife.* Princeton, NJ: Princeton University Press, 2008.

petrels has declined steadily since 1970; only 300 pairs were found in the Antarctic region in 2007.

Human Uses

The continent of Antarctica has no permanent human residents. It neither is owned nor ruled by any single nation. Until the 1959 Antarctic Treaty, which regulates international relationships on the continent, the ice- and snow-covered landmass was surveyed, named, and claimed by multiple nations. As of 2009, the treaty had been signed by forty-seven countries; it sets aside Antarctica as an environmental and scientific preserve.

The peninsula has a rich history of exploration. American Nathanial Palmer (1799–1877) visited the region in 1820 and named it Palmer Peninsula. In 1832, the British renamed a portion of the area Graham Land after Sir James R.G. Graham (1792–1861), First Lord of the Admiralty at the time, and named the entire peninsula Trinity. Argentina claimed the region in 1940 with a name of *Tierra de San Martín* (Land of San Martín). Then in 1942, Chile decided to call the area *Tierra de O'Higgins* (Land of O'Higgins). The Antarctic Treaty picked the more neutral name "Antarctic" for the peninsula, meaning the opposite of the northern arctic.

Today, there are nineteen scientific research stations on the Antarctic Peninsula; these represent the highest concentration of research stations on the continent. Seven facilities are clustered on islands near the tip, including those operated by Argentina, Brazil, Chile, China, Poland, Russia, and Uruguay. While most of these facilities operate year-round, and the largest, the Chilean Marsh/Frei, houses 150 people with the area's only regional airport, no residents stay more than a few months at a time due to the isolated and extreme conditions.

Pollution and Damage

The peninsula's arctic climate slows the breakdown of organic materials to a fraction of the rate of decomposition in a temperate climate. It often takes several hundred years for cellular material to fully decompose in the Antarctic; to do so, it first must rise above the frozen state for more extended periods.

Trash, human waste, chemicals, oils, and other debris often have been discarded on land or in the Southern Ocean due to the high cost of removal by boat or plane. Human activity dates back to the 1800s, when seasonal hunters arrived by ship and processed seals and whales on land. Visible remains from those endeavors still are present on the peninsula, including garbage, collapsed wooden buildings, tools, and even piles of bones. Modern scientific research stations host, on average, 4,000 people during the summer and 1,000 during the winter. Although spread out over the continent, their activities generate substantial waste that impacts the Antarctic environment.

In 1989, the 400-foot (122-meter) Argentine naval ship *Bahia Paraiso* ran aground on the Antarctic Peninsula near Palmer Station. The torn hull leaked 150,500 gallons (569,704 liters) of oil and fuel into a 40-square-mile (104-square-kilometer) area, killing hundreds of seabirds and other marine life. The abandoned ship sank; joint cleanup efforts eventually involved pumping out the remaining fuel from inside the hull.

The practice of releasing untreated human waste into ocean water negatively impacts the biotic community. As nutrient-rich sewage enters the salt water, coliform bacteria flourish. This concentration of nitrogen and phosphorus is toxic to marine life and kills species of plankton, mollusks, crustaceans, and fish. A 1991 addition to the Antarctic Treaty resulted in increased efforts to improve waste management practices. Years of accumulated garbage was removed from the continent or put in improved upland landfill sites, such as lined pits.

The region around the peninsula, although extremely cold, with an average annual air temperature of 14 degrees Fahrenheit (-10 degrees Celsius), is showing signs of climate change. From 1945 to 2004, the average air temperature increased by 4.5 degrees Fahrenheit (2.5 degrees Celsius), which is five times the average global

This picture shows the Princess Elisabeth Antarctic Station, the first and only "zero emission" polar research station. This Belgian station, which can house up to sixteen scientists, began operating in February 2009, and it runs solely on solar and wind energy. *(Benoit Doppagne/AFP/Getty Images)*

warming rate. Simultaneously, the water temperatures recorded in the Southern Ocean have shown a 0.5-degree Fahrenheit (0.3-degree Celsius) increase between 1950 and 2008. These changes have resulted in an expansion of the melting season by three weeks since the 1980s.

The Larsen Ice Shelf sits near the end of the eastern peninsula. One section of ice shelf, spanning 770 square miles (1,994 square kilometers) and up to 656 feet (200 meters) thick, broke away in 1995 and floated into the Weddell Sea, melting into the salt water. A second portion of the shelf, 1,254 square miles (3,248 square kilometers) in size, broke off in 2002, disintegrating in thirty-five days. In total, 80 percent of the 12,000-year-old shelf, spanning an area larger than the size of the state of Rhode Island, has collapsed since the 1990s.

Mitigation and Management

Since the early twentieth century, seven countries have laid claim to portions of the Antarctic Peninsula: the United Kingdom in 1908, New Zealand in 1923, France in 1924, Australia in 1933, Norway in 1939, Argentina in 1943, and Chile in 1943. International tension mounted in the 1950s during the cold war, as possible military uses of the continent as a strategic base were discussed.

As part of the United Nations International Geophysical Year (1957–1958), twelve nations (Argentina, Australia, Belgium, Chile, France, Japan, New Zealand, Norway, South Africa, the Union of Soviet Socialist Republics, the United Kingdom, and the United States) participated in scientific research in Antarctica. At the end of 1958, a proposal to continue using the region as a research site led to a 1959 meeting of the countries' leaders in Washington, D.C., to negotiate the landmark Antarctic Treaty, which oversees the entire region south of 60° latitude (including the islands and ice shelves).

The treaty covers 10 percent of the world's surface area and 10 percent of its oceans. It sets aside all claims to sovereignty, bans military activities, and makes scientific research the cornerstone of any government presence. The treaty enhanced environmental protection for the region by banning nuclear weapons testing and use, prohibiting dumping of radioactive waste, and making any ship or building open to inspection by appointed observers.

By 2000, thirty-three additional countries had signed the treaty as nonvoting members, representing almost 80 percent of the world population. An organization termed the Antarctic Treaty Parties meets annually to govern the region.

Five amendments since 1959 have enhanced the treaty. The 1964 Agreed Measures for the Conservation of Antarctic Fauna and Flora added protection to plants and animals; the 1972 Convention for the Conservation of Antarctic Seals increased efforts to limit seal hunting; the 1980 Convention on the Conservation of Antarctic Marine Living Resources enhanced protections for marine life; the 1988 Convention on the Regulation of Antarctic Mineral Resource Activities regulated the removal of minerals; and the 1991 Protocol on Environmental Protection to the Antarctic Treaty prevented development, such as construction of permanent, livable communities. The oversight parties have regulated tourism activities, addressed environmental damages, controlled ship traffic patterns, and applied protections to threatened or endangered species.

In the twentieth century, the ocean water and air that surround the peninsula experienced an increase in carbon dioxide buildup. The water lacks certain nutrients such as iron, which limits the phytoplankton that digest excess carbon in the water and air. Antarctic researchers theorize that fertilization of the oceans would result in the growth of more phytoplankton to consume the excess carbon.

In 2004, a project from the University of California, Santa Barbara spread a few hundred pounds of iron sulfate in the Southern Ocean off the Antarctic Peninsula at a concentration of 50 parts per trillion, 100 times the normal levels in these cold waters. Massive algae blooms occurred within a few days, spread out over thousands of square miles, and were visible from satellite images. These small plants consumed approximately 60,000 tons of carbon dioxide during six weeks, producing oxygen in the process. When the plant bodies died, they sunk to the bottom of the Southern Ocean, where the carbon will be held for thousands of years. The carbon is only released into the atmosphere when sediments from the bottom are carried to the surface by upwelling currents and a temperature change causes a release of gaseous carbon into the atmosphere. Although the scientific community considered these results successful, an ethical debate has developed over the role that humans should play in manipulating these fragile ecosystems and the potential long-term impacts of such interference.

In 2006, the International Council for Science, a nongovernmental organization representing 117 national scientific bodies and thirty international scientific unions, and the World Meteorological Organization, a specialized agency of the United Nations, orchestrated a fifty-year anniversary event and declared 2007–2008 the International Polar Year. Goals included 220 scientific research projects and a campaign to educate the public on the fragility of the polar environment.

Selected Web Sites

Antarctica Research Stations and Territorial Claims: http://www.lib.utexas.edu/maps/islands_oceans_poles/antarctica_research_station.gif.

Glaciology in the Antarctic: http://pubs.usgs.gov/fs/2005/3055/.

Goddard Space Flight Center, Larsen B Ice Shelf: http://svs.gsfc.nasa.gov/vis/a000000/a002400/a002421/index.html.

History of the Antarctic Treaty: http://www.antarcticanz.govt.nz/downloads/information/infosheets/AntarcticTreaty.pdf.

International Polar Year, 2007–2008: http://www.ipy.org.

Map of Antarctic Peninsula: http://antarcticsun.usap.gov/AntarcticSun/features/images/antmap.jpg.

Secretariat of the Antarctic Treaty: http://www.ats.aq/.

Further Reading

Berkman, Paul Arthur. *Science into Policy: Global Lessons from Antarctica*. San Diego, CA: Academic Press, 2002.

Fox, William. *Terra Antarctica*. Berkeley, CA: Shoemaker and Hoard, 2007.

Knox, George A. *Biology of the Southern Ocean*. Boca Raton, FL: CRC, 2006.

McGonigal, David. *Antarctica: Secrets of the Southern Continent*. Buffalo, NY: Firefly, 2008.

Monteath, Colin. *Antarctica: Beyond the Southern Ocean*. Toronto, Canada: Warwick, 2005.

Myers, Joan. *Wondrous Cold: An Antarctic Journey*. New York: Smithsonian, 2006.

Scott, Jonathan, and Angela Scott. *Antarctica: Exploring a Fragile Eden*. New York: HarperCollins, 2008.

Shirihai, Hadoram. *The Complete Guide to Antarctic Wildlife*. Princeton, NJ: Princeton University Press, 2008.

Smith, Roff. *Life on the Ice*. Washington, DC: National Geographic Society, 2005.

8 North Sea Atlantic Ocean

The North Sea is a partially enclosed body of the Atlantic Ocean located at the edge of northwestern Europe. It is bordered by the United Kingdom on the west, Norway and Denmark on the east, and The Netherlands and Germany on the south. With an area of 222,008 square miles (575,001 square kilometers), the North Sea is 620 miles (998 kilometers) long and 350 miles (563 kilometers) wide. It is the world's thirteenth-largest sea, with an estimated volume of 12,955 cubic miles (54,000 cubic kilometers).

Two main channels link the North Sea to the rest of the Atlantic Ocean: the smaller is the Strait of Dover (between 4 and 20 miles, or 6.4 and 32 kilometers, wide) in the south, between the United Kingdom and France; and the larger opening (330 miles, or 531 kilometers, wide) is in the north between Scotland and Norway, where the North Sea intersects with the Norwegian Sea and the Atlantic Ocean. Compared with other saltwater bodies, the North Sea is shallow, with an average depth of 308 feet (94 meters), and its deepest point is 2,297 feet (700 meters).

The North Sea did not exist until 250 million years ago, when it formed during the Paleozoic Era. At that time, a smaller inland body of freshwater sat against The Netherlands and Germany, surrounded by desert plains to the north and west. The movement of three small continental plates (Avalonia, Baltica, and Laurentia) formed a north–south oriented rift down the middle of the Laurentia Plate. This rift filled with Atlantic Ocean water 100 million years ago. The Avalonia Plate formed the west and south coast, and the Baltica Plate formed the eastern coastline of the new sea. A body of water this size typically sits on an ocean tectonic plate; however, the North Sea is an epicontinental body, because it sits over a portion of the northern European continental shelf.

Approximately 65 million years ago, the general boundaries of the North Sea were much like they are today. Over time, ice ages and the movement of glaciers continued to shape shorelines and ocean shallows by transporting soil and rocks. The movement of glaciers stripped the western coastlines, forming jagged scars, and then deposited the material along southern coasts in thin layers. Additional evidence of this shaping includes three shallow banks to the south that rise out of the water to the surface from the accumulated heavy deposition. These include the Dogger Bank, Great Fisher Bank, and Jutland Bank. What was a deep sea (20,000

Norway's longest sustained export commodity has been stockfish (white fish like cod), such as this catch in the North Sea. After being processed and air-dried for several months in a cool environment, the unsalted fish have a storage life of several years and commonly are exported to markets in northern Europe, Croatia, Italy, and Portugal. (© izzog/Fotolia)

feet, or 6,096 meters, in places) was filled in slowly by sediment. Sedimentary rock formed over decaying organic matter, creating pockets of oil and gas.

The waters of the North Sea vary in salinity. While the North Atlantic's salt content is around 35 parts per thousand, areas in the middle of the North Sea may be 32 parts per thousand, and by the coast it can be as low as 25 parts per thousand, especially in springtime due to freshwater runoff.

The water temperature also varies over a significant range, based on the section of the sea and the season. In the north, water temperatures often drop below 40 degrees Fahrenheit (4.4 degrees Celsius) due to the influx of arctic waters. In the south, water in shallow coastal estuaries can reach 80 degrees Fahrenheit (26.6 degrees Celsius) in the summer months. The sea has a dominant counterclockwise current, and it takes up to two years for waters to make this circuit, moving at a depth of up to 330 feet (101 meters).

The North Sea is known for its rough weather and large waves, which are heavily influenced by the unprotected open north. During one such storm on New Year's Day in 1995, an oil platform measured an 84-foot-high (26-meter-high) wave. In 2000, a ship positioned off Scotland measured a wave at 95 feet (29 meters).

Several countries, such as Denmark, Germany, and The Netherlands, are exposed to North Sea flooding due to extensive coastal lowlands. Flooding has inundated communities and taken hundreds of lives during the last two centuries, resulting in multiple construction efforts, such as dams, dikes, seawalls, and floodwater pumping stations, to hold back the pressing ocean waters. These efforts have been fairly successful in protecting communities susceptible to flooding from the small to medium events, but the largest floods are not halted by current structures.

The cold North Sea currents are plentiful in nutrients that support a diversity of marine species. Shallow waters host thick beds of eelgrass (*Zostera marina*) and kelp forests (*Laminaria hyperborea*) grown in deeper waters. An estimated 230 species of fish live at varying depths. The Atlantic cod (*Gadus morhua*) lives along the ocean bottom and grows up to 6 feet (1.8 meters) long over fifteen years. Other commercially sought species, such as herring (*Clupea harengus*), mackerel (*Scomber scombrus*), pout (*Trisopterus luscus*), prawn (*Pandalus borealis*), sand eel (*Ammodytes marinus*), sole (*Solea solea*), and whiting (*Merlangius merlangus*), also prefer the northern, deeper water.

Upward of 10 million seabirds live in the North Sea region, including twenty-eight species that make up the 4 million individuals that breed along the north coast in the summer. Offshore ocean birds include the northern fulmar (*Fulmarus glacialis*), which spends much of its life at sea feeding on the surface, eating fish, shrimp, squid, and plankton. Other species, such as the coastal

NORTH SEA MAMMALS

Species	Population Estimate
Common Minke Whale (*Balaenoptera acutorostrata*)	20,000
Grey Seal (*Halichoerus grypus*)	60,675
Harbor Porpoise (*Phocoena phocoena*)	268,300
Harbor Seal (*Phoca vitulina*)	33,240
White-beaked Dolphin (*Lagenorhynchus albirostris*)	10,900

Source: Organization for the Protection of the Marine Environment of the North-East Atlantic, 2000.

common tern (*Sterna hirundo*), breed in large colonies on islands where mainland predators cannot reach them.

Mammals living in the North Sea include the 9-foot-long (2.7-meter-long) white-beaked dolphin (*Lagenorhynchus albirostris*). This very social species often feeds near whales and interacts with traveling vessels, darting back and forth across the bow. Whale species that commonly enter the North Sea include the minke whale (*Balaenoptera acutorostrata*), which grows to a length of 23 feet (7 meters) and can weigh 4 to 5 tons. This species relies on a series of large filter plates (called baleen) in its mouth (instead of teeth) with which it captures large volumes of plankton for food. The population of minke whales in the northern Atlantic Ocean was estimated by the International Whaling Commission to be as large as 245,000 in 2009.

Human Uses

The thirteen countries that ring the North Sea are the United Kingdom, Denmark, France, Belgium, The Netherlands, Germany, Poland, Russia, Latvia, Estonia, Finland, Sweden, and Norway. These nations host a coastal population of 17 million residents. The North Sea possesses the world's busiest shipping port, Rotterdam, The Netherlands, which processes 250 million tons of cargo per year. Coastal industries, including fishing, oil and gas extraction, and sand-gravel mining, are cornerstones of the regional economy and culture.

What began back in the fifth century as a way of life that relied upon small boats and handheld lines to fish has grown into an industrial fleet of ships that trawls in deep waters with hydraulic nets. To improve their catch and increase profit, today's vessels are bigger (many are more than 100 feet, or 30 meters, long); use modern food preservation techniques, such as processing and freezing fish once caught; and apply technology such as global positioning systems (GPS)

NORTH SEA COD STOCKS

Year	Estimated Population (Metric Tons)
1963	173,062
1985	91,491
2001	41,887
2004	50,706
2008	54,013

Note: A healthy reproductive fish population (also called a stock), is measured by its weight. In order for the North Sea Cod to recover, the target weight of these fish needs to stay above 150,000 metric tons (165,346 tons), after the commercial catch amount is calculated, for three years in a row.

Source: International Council for the Exploration of the Sea (ICES), 2009.

and side-scan sonar to locate underwater fish. Commercial fish include cod, haddock, hake, herring, mackerel, pout, salmon, tuna, and whiting.

By the 1970s, overfishing threatened several of these species, including cod and herring. During that decade, fishermen supplied 2.6 million tons of fish annually to more than 500 million people. Cod boats landed 300,000 tons of fish annually in the 1980s. By the late 1980s, many commercial fishermen responded to the lack of cod and herring species by changing venues, retrofitting their boats with longlines (series of hooks) instead of nets, and focusing on other species. Some boats went after fish such as the spiny dogfish shark (*Squalus acanthias*), which was abundant. Some traveled farther out to sea and fished for longer periods of time. By 2000, despite attempts to adjust fishing techniques, the cod and herring populations continued to shrink, and the fishing industry's revenue as a whole dropped by over 50 percent in some areas.

In 1965, underwater sonar scans and drilling revealed large oil deposits under the North Sea. By 2006, more than 435 platforms had pumped 3.2 million barrels of oil per day, at its peak rate, through 5,000 miles (8,045 kilometers) of pipeline. By 2008, the United Kingdom alone had pumped 41 billion gallons (155 billion liters) of oil from the North Sea floor; the nation expects to extract another 25 billion gallons (95 billion liters) by 2040. Approximately 50,000 people are employed in the oil industry in the region.

In 1998, the North Sea oil production peaked at 6 million barrels per day and accounted for 9 percent of world oil production. This amount declined over the first decade of the twenty-first century because the remaining oil is more difficult and dangerous, and therefore costly, to extract. Overall, it is estimated that more than 50 percent of the total reserves of North Sea oil had been removed from the undersea terrain as of 2008.

Pollution and Damage

The region's fishing, oil drilling, and mining industries have had major impacts on the North Sea. Overfishing, in particular, has impacted the ecological balance of this ocean environment. For example, removal of too many baitfish, such as mackerel, has meant fewer tuna, which rely on the small fish as a food source. Regional conflict has been waged across the open waters. Fishing disputes arose between Iceland and Great Britain in 1958, 1972, and 1976. All occurred between fishing boats at the opening of the North Sea where it meets the larger Atlantic Ocean. Iceland enforced extensions of their sovereign fishing waters and used navy vessels to keep boats of other nationalities away from Icelandic waters. This prevented British trawlers from accessing viable cod fishing grounds.

Called the "Cod Wars," these disputes sometimes involved military vessels. A negotiated settlement was reached in 1976. Iceland was able to protect fishing up to 200 miles (322 kilometers) offshore, which gave the small country some economic security in the face of steady declines in regional fish catches.

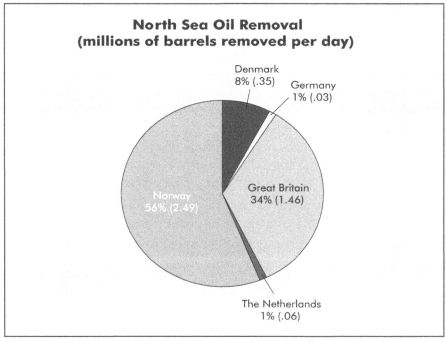

**North Sea Oil Removal
(millions of barrels removed per day)**

Denmark 8% (.35)

Germany 1% (.03)

Norway 56% (2.49)

Great Britain 34% (1.46)

The Netherlands 1% (.06)

Source: Energy Information Administration, 2007.

Oil spills are a particular hazard to the ocean's delicate ecosystem. While large accidents are rare, the cumulative impact of hundreds of small releases has altered the undersea environment. The pumping operations of oil drilling production platforms sometimes release "muddy waste"—a brew of oil, detergents, and petrochemicals that is highly toxic to all living organisms—into the surrounding water. This heavy waste settles to the bottom of the sea, eventually moving with floor currents up to 1,000 feet (305 meters), and it is highly toxic to all living organisms.

The sea also has its labor hazards: A 1988 oil fire caused an explosion on the Piper Alpha Platform (about 120 miles, or 193 kilometers, northeast of Aberdeen, Scotland's third-largest city), killing 167 workers and triggering releases of oil from three dozen interconnected platforms.

Despite the North Sea being fed from the cold-water currents of the Arctic and North Atlantic oceans, there has been an increase in the temperature of the North Sea water due to global climate change. Between 1977 and 2002, the University of East Anglia in Norwich, England, measured a 1.1-degree Fahrenheit (0.61-degree Celsius) increase in average North Sea water temperatures. They surveyed area fish and found that twenty-one of thirty-six fish species responded by moving between 30 and 250 miles (48 and 402 kilometers) into colder, deeper waters.

In 2007, another impact from the increase in water temperature was noted off the coast of Scotland. A 10-square-mile (26-square-kilometer) by 35-foot-deep (10.6-meter-deep) pack of mauve stinger jellyfish (*Pelagia nocticula*) attacked fish populations in November. Known for their mysterious purple glow and painful stings to bathers in the warmer Mediterranean Sea, this species had followed warm waters northward. The jellyfish overwhelmed farmed salmon populations by massing on net pens and caused a $2 million loss.

Up to 60 percent of The Netherlands sits below sea level. Flooding from storm surges has been a fact of life for centuries. Beginning in the Middle Ages, residents opened hand-dug trenches at low tide to remove water from flooded fields. First windmills and then, centuries later, diesel-powered pumps moved millions of gallons of water from inland areas back to the sea.

In 1919, the Dutch government undertook the Zuiderzee, a $5 billion hydraulic engineering project to reclaim lowlands for farming. In 1927, a 19.8-mile-long (31.9-kilometer-long) dike called the Afsluitdijk was built to separate the mainland from the sea near Amsterdam. The drained former seabed sunk approximately 3 feet (0.9 meters) and was planted with salt-tolerant crops.

Although the series of dikes, dams, and pumping stations has remained structurally sound and generally protects residents and land against most storm surges, the project has had a considerable environmental impact. By 1996, up to

Delta Works, a series of dams, sluices, locks, dikes, and storm surge barriers, designed to protect The Netherlands from flooding, was built in response to the North Sea flood of 1953. *Oosterscheldekering* is a 5.59-mile (9-kilometer) storm surger barrier, the largest of the Delta Works structures. *(© Wil Tilroe-otte/Fotolia)*

1,351 square miles (3,499 square kilometers) of saltwater habitat had been drained, including coastal saltwater wetlands that were critical in supporting terrestrial and aquatic species and absorbing floodwaters.

Mitigation and Management

In 1980, Seas At Risk, a European association of nongovernmental environmental organizations working to protect and restore the marine environment of the European seas and the greater northeastern Atlantic region, introduced the North Sea Conference as a forum for bordering countries to work collectively on regional water challenges. Since conference leaders first convened, they have adopted several agreements to manage and protect the resources of the region.

Annual meetings often focus on three of the primary threats to the environment of the North Sea: water quality, which is affected by farm waste, human waste, industrial pollution, storm water runoff, and trash; overfishing; and the impacts of shipping, including marine waste discharges and air pollution. The conference periodically has supported fishing closures and the use of new gear—including, for example, nets with larger holes—to stem overfishing. In addition, conference leaders state that if no measures are taken to curb ship engine emissions, by 2020 ships will generate more air pollution annually than all European land transportation combined.

In response to overfishing, International Council for the Exploration of the Sea (ICES), the world's oldest intergovernmental organization concerned with marine and fisheries science, based in Copenhagen, Denmark, issued a set of reports in the 1990s. The council warned that the cod stock was "severely overexploited" and in danger of a "complete industry collapse." It recommended a scientifically determined set of catch limits to allow the cod to rebuild its natural stocks.

For the first time in history, in 2001 the European Union (EU) imposed Europe-wide catch limits for commercial boats to protect North Sea fish. The catch limit for cod was reduced from 81,000 tons per year to 48,600 tons. In addition, technical guidelines required new types of fishing gear to protect young-er fish.

Despite poor enforcement and an emergent black market, cod stocks showed a small improvement in 2007. In 2008, ICES advised fishing industry leaders to cut their North Sea catches even further to enable populations to recover. In an effort to facilitate recovery, EU regulators again lowered annual cod catch limits to 22,000 tons until 2009.

Samplings of cod populations taken late in 2009 produced encouraging re-sults: Some areas have experienced a 42 percent climb in fish. Scientists hope that within a few years, the population will grow large enough to permit the fishing fleet to resume widespread fishing across the North Sea.

Selected Web Sites

International Council for the Exploration of the Sea: http://www.ices.dk.
North Sea Commission: http://www.northseacommission.info/.
The North Sea Region Programme: http://www.northsearegion.eu.
Seas at Risk, North Sea Conference Documents: http://www.seas-at-risk.org/n2.php?page=9.
The United Kingdom Offshore Oil and Gas Industry Association: http://www.oilandgas.org.uk.

Further Reading

Barale, Vittorio, and Martin Gade, eds. *Remote Sensing of the European Seas.* London: Springer, 2008.

Brand, Hanno, and Lee Muller, eds. *The Dynamics of Economic Culture in the North Sea and Baltic Region.* Hilversum, The Netherlands: Uitgeverij Verloren, 2007.

Finch, Simon, Kenneth Thomson, and Vincent Gaffney, eds. *Mapping Doggerland: The Mesolithic Landscapes of the Southern North Sea.* Oxford, UK: Archaeopress, 2007.

Ilyina, Tatjana P. *The Fate of Persistent Organic Pollutants in the North Sea.* New York: Springer, 2007.

International Council for the Exploration of the Sea. *Report on the State of the Stock: Cod in the North Sea.* Copenhagen, Denmark, 2006.

Kirby, Alex. "North Sea Cod 'Face Commercial End.'" *BBC News,* December 16, 2002.

North Sea Conference. *The Environmental Impact of Shipping and Fisheries.* Gotenborg, Sweden: North Sea Conference, 2006.

Organization for the Protection of the Marine Environment of the Northeast Atlantic. *Status Report for the Greater North Sea.* London, UK: OSPAR, 2000.

9 Hawaiian-Emperor Seamount Pacific Ocean

The Hawaiian-Emperor Seamount is a prominent line of underwater mountains in the north-central Pacific Ocean that stretches 4,300 miles (6,919 kilometers) from the Aleutian Islands south to the Hawaiian Islands. Cutting through three Pacific basins (Northwest, Northeast, and Central), the seamount, also known as a ridge, is built from at least 125 large volcanoes and hundreds of smaller peaks. The mountains are up to 30 miles (48 kilometers) wide, and many soar 20,000 feet (6,096 meters) from the ocean bottom. Key islands in this undersea chain include the Hawaiian Islands, the Midway Atoll, and several smaller uninhabited reefs.

The geologic history of the Hawaiian-Emperor Seamount dates back 81 million years. A volcanic vein called a hot spot emerged from beneath the Pacific Ocean crust, creating the first mountains. As the Pacific tectonic plate moved in a northwestern direction, the hot spot stayed in the same place, creating a line of volcanoes.

At 400,000 years old, the Hawaiian Islands are the youngest formation of the hot spot, and, as demonstrated by ongoing multiple eruptions, the ridge is growing in size. The Hawaiian-Emperor Seamount contains at least 186,000 cubic miles (775,282 cubic kilometers) of lava, enough to cover the state of California with a mile-thick layer of rock.

The oldest section of the seamount is attached to the Asian tectonic plate near Russia and Alaska. The Asian Plate was traveling in a northern direction until 42 million years ago, when its collision with the Indian Plate caused the Pacific Plate to turn almost 60 degrees in a more northwestern direction, forming a distinct L shape across the north-central Pacific Ocean.

KEY FEATURES OF HAWAIIAN-EMPEROR SEAMOUNT

Name	Belongs to Ridge/Seamount	Age (Years)
Suiko Seamount	Emperor Seamount	59.6 million
Nintoku Seamount	Emperor Seamount	56.2 million
Ojin Seamount	Emperor Seamount	55.2 million
Midway Island	Hawaiian Ridge	27.7 million
La Perouse Pinnacle	Hawaiian Ridge	12 million
Hawaiian Islands	Hawaiian Ridge	400,000

The formation of a hot spot volcano takes millions of years and involves several stages. The first part begins deep under the ocean crust. Pressurized lava finds a weak spot where the ocean floor is thin. Lava presses against a crack in this material and eventually flows through the fracture. The magma is low in volume and, due to the extremely high water pressure at the bottom of the ocean, forms rock in layers that resemble pillow shapes. Over years, the tectonic plate moves, and the hot spot stays in the same location. This creates a long line of mountain formations known as seamounts.

Once the hot spot volcano breaks the surface of the water, numerous underwater explosions occur from the oxygen in the atmosphere interacting with molten lava. Landslides of new rock are common as the mass stabilizes. This process may continue for up to 250,000 years. If the volcano breaks the surface of the ocean, rain, wind, and waves will heavily erode the new landform.

Most hot spot eruptions do not form into islands but instead form underwater ridges such as the Hawaiian-Emperor Seamount. This is because the intensity of the eruption, combined with the movement of the tectonic plate, keeps the seamounts smaller in size.

This seamount provides an exceptional habitat for marine life. In the deep open ocean, few prominent underwater landforms exist. Seamounts alter the movement of water, creating eddies and upwellings where nutrients are delivered to species along the ridge. Pink coral (*Corallium secundum*) and red coral (*Corallium lauuense*) live at depths up to 1,800 feet (549 meters). Once they are established on seamount slopes, crabs, fish, sea urchins, snails, sponges, and starfish appear.

The Hawaiian monk seal (*Monachus schauinslandi*) forages along the seamounts, diving to 2,000 feet (610 meters) in search of small fish, octopus, and spiny lobster. This light gray mammal grows up to 8 feet (2.4 meters) long and weighs as much as 600 pounds (272 kilograms). Environmentalists estimate that, due to human disturbances such as fishing, around 1,300 of these endangered seals survive today.

Green sea turtles (*Chelonia mydas*) forage up and down the seamount among thick sea grass, kelp, and corals. They feed strictly on plants, and their young spend their first three years in the deep nutrient-rich waters.

A number of ocean birds occupy the open waters above the seamounts. The Buller's shearwater (*Puffinus bulleri*) spends almost its entire life in the open ocean away from land. Feeding at the surface, its diet consists of small fish, krill, and jellyfish. Up to 4,000 of these large seabirds die each year from getting caught in fishing nets and lines.

The Laysan albatross (*Phoebastria immutabilis*), a large seabird with an 8-foot (2.4-meter) wingspan, lives in the north Pacific. With a population in the Hawaiian Islands that exceeds 2.5 million, these birds move from the open ocean to secluded islands to reproduce. Their courtship practices include elaborate dances with at least twenty-five different movements exhibited by both sexes.

The sea floor around the Hawaiian Islands is 4 miles (6.4 kilometers) deep, pitch black, and under intense water pressure. Although the slopes of the island ridge host many species, the offshore Pacific bottom is less species friendly. In water up to 20,000 feet (6,096 meters) deep, fewer animals exist, and they must rely on less food. Most species survive on specks of organic material that drift down from the surface.

Whale species, such as the humpback whale (*Megaptera novaeangliae*), arrive annually from the Arctic Sea to birth their young in the warm waters. More than 1,100 whales settle here each November and stay until May.

Hawaiian monk seals (*Monachus schauinslandi*) live between twenty-five and thirty years and average about 400 pounds (181 kilograms). The name "monk" reflects the fact that these animals live a relatively solitary existence as compared to other species of seals, which tend to gather in large colonies. (© *Kathy L/Fotolia*)

Some humpbacks, however, die in these Pacific waters. When a 50-foot-long (15-meter-long), 90,000-pound (40,823-kilogram) carcass sinks to the bottom of the ocean, practically a new habitat is formed. First, sharks and fish feed on the flesh that is composed of as much as 60 percent fat. Up to thirty-eight species of fish feed on the carcass for the first year, including the hagfish (*Eptatretus stoutii*), cleaning the meat from the large bones. Smaller creatures, such as bone-eating worms (*Osedax frankpressi*), cover the skeleton and consume the remaining oils and fat.

After several years, tons of the whale's bones remain but little oxygen is present, so marine species that live on anaerobic processes take over, including mussels, giant clams, and other crustaceans. In all, as many as 4,400 species may be involved in the final consumption of the world's largest mammal.

Human Uses

Except for a few small islands, the Hawaiian-Emperor Seamount sits underwater. Although numerous landmasses may have existed millions of years ago, almost all have eroded below the surface.

The northernmost habitable island along the edge of the seamount is the Midway Atoll (these coral-rich islands feature a ring of land and a sunken middle filled with salt water). This circular barrier reef is 2.4 square miles (6.2 square kilometers) in area and contains three small islands (Eastern, Sand, and Spit). As the name suggests, the Midway Atoll is located halfway between the United States and Asia, roughly 2,800 miles (4,505 kilometers) from San Francisco, California, and 2,500 miles (4,023 kilometers) from Tokyo, Japan.

The atoll is a territory of the United States and was occupied by fewer than sixty residents in 2009. Once a shield volcano, a 516-foot-thick (157-meter-thick) reef encircles the collapsed mountaintop today. This atoll was a strategic island for the U.S. Navy during World War II, but it has since become inactive as a military site.

The Hawaiian Islands sit 1,250 miles (2,011 kilometers) southeast of the Midway Atoll. Nineteen major islands and 118 smaller atolls cover 6,423 square miles (16,636 square kilometers) of ocean. The North American continent is 1,860 miles (2,993 kilometers) away.

The biggest island, Hawaii, is 93 miles (150 kilometers) in diameter and hosts 62 percent of the total land area of all the islands. It contains the highest peak, Mauna Kea (meaning "white mountain" in the Hawaiian language, referring to its snow-covered winter summit), which is 13,796 feet (4,205 meters) high. If above- and below-water elevations are combined, this is the tallest mountain on the planet. It rises some 19,028 feet (5,800 meters) from the sea floor to reach

HAWAIIAN FISHING STATISTICS
(Annual Pounds Caught)

Fish	2001	2005
Bigeye Tuna (*Thunnus obesus*)	4,674,742	11,788,857
Yellowfin Tuna (*Thunnus albacares*)	3,884,851	3,249,497
Albacore Tuna (*Thunnus alalunga*)	2,850,567	1,033,073
Swordfish (*Xiphius gladias*)	313,535	3,599,151
Sharks (*superorder Selachimorpha*)	145,765	441,007
Spiny Lobster (*Panulirus marginatus*)	7,813	11,864

Source: National Marine Fisheries Service, Pacific Islands Fisheries Science Center, 2007.

a total height of 32,824 feet (10,005 meters). The island of Hawaii is built from five volcanoes. Two of them, Mauna Loa and Kilauea, are active. Between 1983 and 2002, lava flows added 543 acres (220 hectares) of igneous rock to the island.

The human population on the eight islands totaled 1.3 million in 2009. Although the Hawaiian Islands have developed a diversified economy that includes agriculture, manufacturing, and tourism, a significant segment still relies on fishing.

The Hawaiian fishing industry ranks as the sixth largest in the United States, with $69.7 million in earnings in 2006. That year, 30 million fish were caught by five methods: bottom fishing using a hook and line (for grouper and snapper), long-line fishing (for tuna and swordfish), surface trolling (for mahi mahi, marlin, and wahoo), net fishing (for scad), and using traps (for lobster). The average Hawaiian eats 90 pounds (41 kilograms) of fish per year, more than twice the continental U.S. resident's average of 40 pounds (18 kilograms).

Pollution and Damage

Only the southern end of the Hawaiian-Emperor Seamount is volcanically active, hosting numerous earthquakes and eruptions each year. During 2006, the region hosted five earthquakes greater than 4.0 magnitude; the highest measured 4.7. The strongest was underwater at the Loihi Seamount, 3,180 feet (969 meters) below sea level. These events resulted in magma releases, explosions, and rock slides. This dynamically active region is in the underwater mountain-formation stage, as several of the islands and underwater ridges are in the process of stabilizing and growing in size.

During World War II, the Pacific Ocean battles between Japan and the United States took a toll on the northern Pacific Ocean. Midway Island and the Hawaiian Islands (including Pearl Harbor) were heavily polluted. Midway Atoll was turned into a battleground for three days in June 1942, when two battleships, five aircraft carriers, and 328 aircraft were sunk in the region. These sunken warships and planes, loaded with bombs, chemicals, and fuel, leached into the ocean. More than 3,000 soldiers living on Midway for several years created tons of human waste and trash, which they dumped in the ocean due to the lack of suitable upland.

Years of fishing industry growth along the Hawaiian ridge have resulted in depleted fish populations. In 1991, a fleet of long-liners deployed approximately 12 million hooks; by 2006, the fleet had surged to 33 million hooks.

In addition to affecting fish populations, such large-scale fishing has impacted marine animals. The monk seal population in 2007 numbered 1,300 individuals, a precipitous loss from more than 200,000 in 1960. The green sea turtle was estimated at 200,000 individuals in 2007, down from tens of millions in 1950. Much of this loss is due to three factors: disturbed upland habitat from coastal development, entanglements in fishing nets and lines, and reduced food supply from overfishing.

Mitigation and Management

Since 1924, the U.S. Geological Survey has operated the Hawaiian Volcano Observatory on Kilauea to monitor the island's four active volcanoes, Haleakala, Hualalai, Kilauea, and Mauna Loa, and the surrounding area. The observatory provides regular updates to the public regarding geologic activity.

The frequency of earthquakes along the Hawaiian-Emperor Seamount prompted the U.S. government to install the Pacific Tsunami Warning Center in 1968 to protect residents from sudden, damaging waves. The system relies on a network of buoys installed with sensors that monitor the ocean bottom and surface water conditions and transmit data via satellite. An international collaboration of twenty-six nations support the Pacific Tsunami Warning Center, as well as several other facilities capable of tracking tsunamis across the Pacific.

In order to conserve the Hawaiian-Emperor Seamount and its oceanic habitat, American President Theodore Roosevelt created the Hawaiian Islands Reservation in 1909 to halt the overharvesting of seabirds such as the albatross for their feathers.

Although the Midway Atoll served as a stopover for planes and boats and as a U.S. military outpost during World War II and the Korean, Vietnam, and cold wars, the U.S. Navy relinquished its oversight to the U.S. Department of the

Fifty tons of floating plastic arrive at Midway Atoll in the North Pacific each year, brought there by oceanic currents. About 80 percent of this marine trash comes from land, usually washed by rain off highways, open-air landfills, and city streets, into streams and rivers, and then out to sea. The rest generally comes from ship trash and containers lost at sea during storms. *(Daisy Gilardini/The Image Bank/Getty Images)*

Interior in 1988. Subsequently, the atoll was declared the Midway Atoll National Wildlife Refuge in 1998.

Responding to the collapse of the seal and turtle populations throughout the seamount, in 2000 the U.S. federal government intervened to protect both of these endangered species, as well as overexploited fish populations. Regulators occasionally closed the lobster and swordfish seasons and required all Hawaiian longline fishing boats to work with independent observers, who monitored catches and ensured that fishermen complied with the government's strict fishing limits. The new regulations resulted in fishing industry earnings dropping from $88 million to $30 million between 1995 and 2003, but the new rules provided a reprieve for seals, turtles, and other affected species.

In 2004, authorities reopened the swordfish season after new gear was developed that reduced the incidence of turtle deaths. Vessels fished under the explicit state and federal regulation that if seventeen sea turtles and sixteen loggerhead turtles each were caught by Hawaiian commercial boats, the fishing season would end.

The need for a more comprehensive protection for the Hawaiian Islands and surrounding water habitat led to the U.S. federal government's creation of the Northwestern Hawaiian Islands Marine National Monument in 2006. Named the Papahā naumokuā kea Marine National Monument by Hawaiians to honor the local goddess who, tradition holds, gave birth to the islands, this 140,000-square-mile (362,600-square-kilometer) preserve is larger than all of the U.S. mainland national parks combined and comprises the largest marine conservation area in the world.

The monument includes ten islands and atolls, including Midway, and hosts 10 percent of the total U.S. tropical shallow-water reef habitat. Recreational fishing is banned here, and commercial fishing in the area is scheduled to be phased out by 2011 to protect the approximately 7,000 species that live in this unique marine environment.

Selected Web Sites

Hawaiian Volcano Observatory: http://hvo.wr.usgs.gov.

National Oceanic and Atmospheric Administration, Ocean Explorer, Precious Corals in the Hawaiian Archipelago: http://oceanexplorer.noaa.gov/explorations/03nwhi/missions/leg2_summary/leg2_summary.html.

Midway Atoll Wildlife Refuge: http://www.fws.gov/midway/index.html.

Pacific Tsunami Warning Center: http://www.prh.noaa.gov/pr/ptwc/.

Papahā naumokuā kea Marine National Monument: http://hawaiireef.noaa.gov.

U.S. Geological Survey, The Long Trail of the Hawaiian Hot Spot: http://pubs.usgs.gov/gip/dynamic/Hawaiian.html.

Further Reading

Hekinian, Roger, et al., eds. *Ocean Hotspots: Intraplate Submarine Magnetism and Tectonism.* New York: Springer, 2004.

Kim, Karl. *Estimating the Impacts of Banning Commercial Bottomfish Fishing in the Northwestern Hawaiian Islands.* Honolulu: University of Hawaii, 2006.

Miller, Frederic, Agnes Vandome, and John McBrewster, eds. *Hawaii Hotspot.* Beau Bassin, Mauritius: Alphascript, 2009.

Morato, Telmo, and Daniel Pauly, eds. *Seamounts: Biodiversity and Fisheries.* Vancouver, Canada: University of British Columbia, Fisheries Centre, 2004.

Nouvian, Claire. *The Deep: The Extraordinary Creatures of the Abyss.* Chicago: University of Chicago Press, 2007.

Pitcher, Tony J., et al., eds. *Seamounts: Ecology, Fisheries, and Conservation.* Malden, MA: Wiley-Blackwell, 2007.

Rauzon, Mark J. *Isles of Refuge: Wildlife and History of the Northwestern Hawaiian Islands.* Honolulu: University of Hawaii Press, 2001.

Rogers, Alex D. *The Biology, Ecology, and Vulnerability of Seamount Communities.* Gland, Switzerland: World Conservation Union, 2004.

Takahashi, Eiichi, et al., eds. *Hawaiian Volcanoes: Deep Underwater Perspectives.* Washington, DC: American Geophysical Union, 2002.

10 | Sargasso Sea
Atlantic Ocean

The Sargasso Sea is part of the North Atlantic Ocean, located between 20° and 35° north latitude and 30° and 70° west longitude. An elliptically shaped body of water, the sea is approximately 1.4 million square miles (3.6 million square kilometers) in size, stretching 700 miles (1,126 kilometers) north to south and 2,100 miles (3,379 kilometers) west to east. It sits near the continental shelf of the southeastern United States and Caribbean Sea on the west and south and in the open Atlantic Ocean on the north and east. The Sargasso is unique in its microclimate and relative isolation inside the larger Atlantic Ocean, where it is far from landmasses and bordered by ocean currents.

Multiple ocean currents demarcate the Sargasso Sea. These include the Gulf Stream, Northern Equatorial Current, Caribbean Current, Florida Current, and Antilles Current. Pushing nearly 70 million cubic feet (2 million cubic meters) of water every second, these currents create a large circular motion called the North Atlantic Gyre, which rotates clockwise around a section of the northern Atlantic Ocean.

Similar to the eye of a large storm, the center of the North Atlantic Gyre hosts calmer water and weather conditions. Within this often tranquil area sits the Sargasso Sea. Due to the pressure of the rotating current, its water levels are 3 feet (0.9 meters) higher than that of the coastal regions. The Sargasso's water also is known for its salty conditions, warmth, clarity, and deep blue color. While most of the Atlantic has a salt content of approximately 35 parts per thousand, the Sargasso often has a content of 37 parts per thousand. The high salinity is due to a combination of reduced circulation in the center, solar-heated water and air, and low freshwater input.

Sargasso Sea Currents

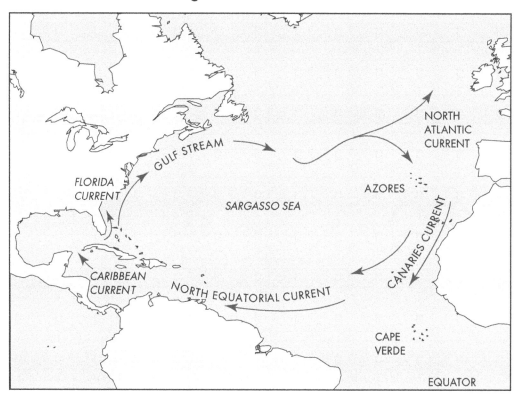

The Sargasso has two noteworthy geologic features. The first, on the west and south, is the edge of the continental shelf of North America. Where the shelf ends, ocean depths drop from approximately 600 feet (183 meters) to almost 5,000 feet (1,524 meters) within a few miles. This marks the intersection of the tectonic plates from the Atlantic Ocean and North America.

The second noteworthy formation is the almost featureless ocean bottom, called the Nares Abyssal Plain, which ranges from 5,000 to 23,000 feet (1,524 to 7,010 meters) deep. This mostly flat region has few ridgelines and is coated with a thick layer of fine red clay, accumulated from years of slow deposition from North American river runoff.

Fifteenth-century Portuguese sailors named the Sargasso after they observed vegetation resembling little grapes, or *sargaço,* floating at the surface in large mats. Two species of seaweed called sargassum, *Sargassum fluitans* and *Sargassum natans,* form a floating vegetative habitat for many species that live in this region. The plants reproduce asexually, breaking off shoots to form new individual plants.

The sargassum fish (*Histrio histrio*) lives among the plants and grows up to 8 inches (20 centimeters) in length. Its color and appendages have evolved to look like parts of the green plants and allow the fish to blend into the seaweed environment.

Camouflaged by seaweed, the sargassum fish *(Histrio histrio)* lies in wait to ambush prey, such as shrimp and small fish. To escape its own underwater predators, the sargassum fish can jump onto floating seaweed for a minute or so, and then bounce back into the water. *(Wolcott Henry/National Geographic/Getty Images)*

Several species of sea anemone (*Actiniaria*) also live within the seaweed, using hidden stingers to paralyze passing prey. Ranging in length from 0.5 inch (1.3 centimeters) to almost 1 foot (0.3 meters), these predators lie in wait for unsuspecting fish.

Small sargassum swimming crabs (*Portunus sayi*) have developed a unique skill to navigate the seaweed beds. This species uses clusters of its own air bubbles to float within the seaweed mat and has a small wing-like flipper on its rear feet that helps it travel through the water column (layers of water between the bottom and the surface that often are defined by chemical or thermal characteristics).

The Sargasso Sea supports two unique species of eel. The European eel (*Anguilla anguilla*) and the American eel (*Anguilla rostrata*) are born in February in the warm waters at the surface of the Sargasso. The tiny eels, known as elvers, float in the slow-moving currents in a northward direction out of the sea to coastal areas either in the eastern United States or western Europe. These eels spend the bulk of their lives in estuaries, where freshwater and salt water mix, continually growing in size. Once they are sexually mature, they swim several thousand miles back to the Sargasso Sea, a journey that takes about eighty days, where they mate in depths around 3,900 feet (1,189 meters).

Human Uses

The Genoese explorer Christopher Columbus sailed through the Sargasso Sea in 1492. When he arrived in dense stands of seaweed, he feared that rocky shoals were nearby, but when the crew took soundings to seek the depth, they did not reach bottom.

In the 1600s, Spanish sailors became stranded in the windless Sargasso waters and were forced to eat or jettison their horses. Thus, the region became known as the "horse latitudes."

Some sailing charts still are marked with warnings about these conditions. The suffering of being stuck in a motionless sea was captured in the English poet Samuel Taylor Coleridge's epic poem *The Rime of the Ancient Mariner* (1798): "Water, water, everywhere, and all the boards did shrink; Water, water everywhere, nor any drop to drink."

The only landmass inside the Sargasso is the Bermuda Islands, a group of approximately 300 islands, of which some twenty are inhabited. Tucked in the sea's northwest corner, they are spread over 20 square miles (52 square kilometers). The principal island, Bermuda Island, is a British territory, located 640 miles (1,030 kilometers) southeast of Cape Hatteras, North Carolina. It was visited often in the sixteenth century by explorers but settled permanently by British sailors who wrecked upon its reefs in 1609.

As of 2009, the Bermuda Islands had a total population of 67,837. Bermudians, per capita, made $91,477 per year in 2008, which is extraordinarily wealthy for a remote island community. Bermuda features low offshore tax rates on personal and corporate income, resulting in the registration of more than 12,500 foreign companies on the island, as well as many corporate employees who call Bermuda home. Economic development comes mainly from tourism (more than 650,000 people visited in 2008) and the financial services industry.

Islanders rely on rainfall for drinking water. Much of the land used to be covered with Bermuda cedar (*Juniperus bermudiana*), but overcutting in the 1830s, disease in the 1940s, and the construction of new buildings have limited coverage today. Bermuda's position in the heart of the Gulf Stream provides year-round tropical warmth. The island's terrestrial environment includes 1,100 species of insects, 750 types of fungi, 360 species of birds, and 167 species of plants. The adjacent coral reefs host hundreds of species of plankton, 286 species of fish, thirty-eight species of corals, thirty-eight types of crustaceans, twelve species of mammals, and other marine life.

The Bermuda fishing industry employs less than 1 percent of the population (610 residents) but earned almost $7.5 million in 2008. There are two environs in which to fish: shallow reef communities and deeper offshore waters. While the coral reef waters host yellowfin grouper (*Mycteroperca venenosa*), stoplight parrotfish (*Sparisoma viride*), and gray or mangrove snapper (*Lutjanus griseus*), the offshore waters contain pompano dolphin fish (*Oryphaena equiselis*), Atlantic blue marlin (*Makaira nigricans*), tiger sharks (*Galeocerdo cuvier*), Atlantic bluefin tuna (*Thunnus thynnus*), and wahoo (*Acanthocybium solandri*). The historic equipment for near-island fishing is the fish pot, a rectangular metal cage with a funnel-shaped entrance. Tackle rod and reels or nets of various sizes are used for offshore fishing.

The Sargasso Sea contains a section of the mythical Bermuda Triangle, or the "Sea of Lost Ships." Located in the western part of the North Atlantic Ocean between Bermuda, the Florida Straits, and the Bahamas, this region is where a number of boats and aircraft have mysteriously crashed or sunk since the 1700s. Up to 1,600 accounts, including stories of boats navigating in circles, crews losing their sanity, and nighttime landscapes illuminated by bright lights, have fed folklore.

Explanations for these events include human navigation mistakes in open water with no land bearings; magnetic irregularities that resulted in abnormal compass readings; mechanical failure of engines or equipment; and harsh environmental conditions here, such as sudden, violent storms, rogue waves, high heat, and low winds. No single reason for the Bermuda Triangle has been scientifically proven, adding mystery and controversy to the surrounding area.

Pollution and Damage

Commercial reef and offshore fishing in the northwestern Sargasso Sea and Bermuda area has resulted in a depletion of fish stocks in the last century. In 1950, two-thirds of all fish eaten on the islands were locally caught. Human population and tourism growth over the next three decades demanded more seafood than boats could supply, and overfishing resulted.

To track the decline of the fish population, the Bermuda Islands government passed the 1972 Fisheries Act, which required local fishermen to submit information on the weights and types of fish they caught. Despite these regulations, certain fish species, such as grouper and snapper, were almost exterminated as the total annual fish catch rose dramatically. In 1975 reported catch amounts totaled 936,964 pounds (425,000 kilograms) of fish; by 1987, this number ballooned to 1.4 million pounds (0.6 million kilograms).

As fish species began to vanish, the government imposed further restrictions when it passed the 1990 Fish Pot Ban, which strictly licensed catch limits in the southern waters off the islands. The fishing ban resulted in the shrinkage of the industry and, consequently, lower catch amounts: 771,617 pounds (350,000 kilograms) in 1992 and 630,522 pounds (286,000) in 2005.

Another contributor to the depletion of the fish population in Bermuda has been pollution of the environment. This mainly has been the result of atmospheric deposition, dredging, eutrophication, and sewage disposal.

Although limited oil drilling is performed in the Sargasso Sea—due to few oil discoveries and general inaccessibility to the ocean floor because of the sea's deep waters—oil pollution is present. Five large oil spills across the North Atlantic

LARGEST OIL SPILLS IN THE NORTH ATLANTIC OCEAN

Year	Ship Name	Location	Spill Amount (Tons)
1967	Torrey Canyon	Scilly Isles, Northeastern Atlantic Ocean	129,000
1978	Amoco Cadiz	Off Brittany, France, Northeastern Atlantic Ocean	253,000
1979	IXTOC-1	Gulf of Mexico	551,000
1993	Braer	Off Shetland, Scotland, North Sea	93,000
2002	Prestige	Off Galicia, Spain, Northeastern Atlantic	67,000
Total:			1,093,000

Source: Oil Spill Case Histories, National Oceanic and Atmospheric Administration, 1967 to 2005.

since 1967 have resulted in the release of more than 1 million tons of oil (twenty-five times the amount of oil released by the 1989 *Exxon Valdez* disaster off the coast of Alaska).

Since oil does not mix with water, it is transported by winds and currents. The lighter components of oil evaporate, but the heavier parts, from unprocessed crude oil, remain in the open water for years. Drifting oil sometimes combines with sediment and trash to form floating tar balls. Scientific research expeditions to the Sargasso Sea regularly find this spilled oil residue and debris when pulling up fine nets or sampling equipment.

Mitigation and Management

Before 1900, the Sargasso Sea was a mystery to both sailors, who found little wind when crossing the area, and scientists, who concluded, without the use of modern analytical equipment, that it was devoid of life. Scientists, who saw few marine species, believed that low iron content in the salt water, intermittent circulation, warm weather, and a mostly flat sea floor meant that the Sargasso Sea was inhospitable to ocean life.

This assumption began to change in the 1990s when National Aeronautics and Space Administration (NASA) satellites recorded images of nutrient-rich eddies (circular currents) breaking off from the Gulf Stream and swirling into the Sargasso Sea. Later, scientists at the Woods Hole Oceanographic Institute discovered that these currents were supplying waters with nutrients from the sea floor, including nitrogen and phosphorus. This influx of nutrients resulted in massive blooms of phytoplankton on the surface, up to 100,000 times the average amount of phytoplankton growing elsewhere in the region.

Specifically, three research projects undertaken in the 1990s and 2000s provided evidence of a rich undersea environment in the Sargasso Sea. The first, the Bermuda Atlantic Time-Series Study (BATS)—a long-term study established in 1954 with a research site located in the open ocean 50 miles (80 kilometers) southeast of Bermuda—has provided decades of data about the region. Each month, water samples are collected from several different depths of the water column to measure chemical and biological contents. Based on these samples, scientists have written hundreds of research papers on topics as varied as phytoplankton growth rates and vertical nitrate flux and have gained a better understanding of the Sargasso environment and the changes occurring within it.

In 2004, a scientific study by the Institute for Genomic Research (known as the J. Craig Venter Institute since 2006), the University of Southern California, and the Bermuda Institute of Ocean Sciences inventoried the DNA of plankton, the most abundant form of sea life in the Sargasso. The study found that the sea's marine life was more prolific than researchers previously had believed. Five 50-gallon (189-liter) containers that sampled surface water held 1,800 species of plankton. Approximately 782 of these microbial organisms, collectively made up of an average of 1.2 million different genes (humans have just over 26,000), contained photoreceptor genes. A photoreceptor is a unique cell that is able to convert light into energy; fewer than 200 photoreceptor species previously were documented planetwide. With limited nutrients available, hundreds of these organisms have evolved to rely on a plentiful source of energy: sunlight.

FACTS ABOUT SARGASSO SEA EDDIES

- They can contain up to 1,200 cubic miles (5,002 cubic kilometers) of water.
- Once splintered from the Gulf Stream, currents move at up to 3.5 miles per hour (5.6 kilometers per hour).
- These colder rings of water can span 200 miles (322 kilometers) in diameter.
- Located near the Gulf Stream, some eddies last up to two years before losing momentum.

Source: Eddies Dynamics, Mixing, Export, and Species Composition Project, Woods Hole Oceanographic Institute, 2007.

The third scientific effort that better chronicled the life in the Sargasso Sea was part of an international project called the Census of Marine Life. In 2006, a research ship with scientists from fourteen countries traveled to the sea to do species inventories. The project found more than 1,000 organisms at a total of five sampling sites. There were hundreds of species of shrimp-like copepods, sixty-five species of ostracods (a shrimp cousin), twenty-four species of pteropods (swimming snails), and 120 species of fish, including blue angelfish (*Holacanthus bermudensis*), black dragonfish (*Idiacanthus atlanticus*), and oarfish (*Regalecus glesne*).

The data gathered and its analysis have answered many questions about the sea's unique weather and water conditions and the distinctive species living within its borders. With phenomena such as climate change and ocean acidification prevalent throughout the global oceans, this research is critical for the thoughtful management and oversight of the so-called "sea with no shores."

Selected Web Sites

Bermuda Atlantic Time-Series Study: http://bats.bios.edu.

Gulf Stream: http://www.gma.org/space1/eddy.html.

Image Quest Marine, Sargassum Images: http://www.imagequest3d.com/photos/sargassum/index.htm.

Sargasso Zooplankton: http://www.cmarz.org/CMarZ_RHBrown_April06/images/animal_photos/cmarzgallery_animals.html.

United Nations Educational, Scientific and Cultural Organization, Bermuda Overview: http://www.unesco.org/csi/pub/papers/smith.htm.

Windows to the Universe, Sargasso Sea: http://www.windows.ucar.edu/tour/link=/earth/images/SargassoSea_image.html.

Further Reading

Coleridge, Samuel Taylor. *The Rime of the Ancient Mariner.* 1798. Edison, NJ: Chartwell, 2008.

Gruber, Nicolas, and Charles Keeling. *Seasonal Carbon Cycling in the Sargasso Sea Near Bermuda.* Berkeley, CA: University of California Press, 2001.

Longhurst, Alan R. *Ecological Geography of the Sea.* Burlington, MA: Academic Press, 2007.

Pauly, Daniel, and Jay Maclean. *In a Perfect Ocean: The State of Fisheries and Ecosystems in the North Atlantic Ocean.* Washington, DC: Island, 2003.

Pollack, Andrew. "A New Kind of Genomics, with an Eye on Ecosystems." *The New York Times,* October 21, 2003.

Smith, Struan. *Bermuda: Environment and Development in Coastal Regions and in Small Islands.* St. George, Bermuda: Bermuda Biological Station for Research, 2000.

OCEANS
CONCLUSION

11 | State of the Oceans

The World Ocean plays a critical role in maintaining the planet's ecosystems and is essential to human health and well-being. Oceans help supply essential moisture to provide precipitation for freshwater supplies and support plant and animal populations. They also provide high-protein food resources and additional agricultural benefits, from animal feed to fertilizers.

Yet an increasingly populated world and more than a century worth of human impacts—overfishing, oil extraction and spills, pollution from land sources, coastal development, and humankind's lack of understanding of the complexity of the ocean ecosystem—have led to an ecologically challenged ocean environment.

Human Impacts to the Oceans

Up until the twentieth century, the human population had only minor impacts on the ocean environment. However, over the past 100 years, rapid acceleration of growth has affected ocean habitats worldwide.

By 1800, the world population had reached 978 million. By 2009, it was 6.7 billion, and it is estimated to reach 9 billion by 2050. This human expansion has relied in part on resource consumption from the oceans. After a century of exhaustive removals of marine life and oil worldwide, there is a reduced diversity and quantity of aquatic species, as well as generally degraded seawater quality.

Overfishing

In 2007, the United Nations concluded that up to 40 percent of the ocean fisheries worldwide has been heavily impacted from overfishing. Additionally, 80 percent of marine life is at risk from overfishing, and only 4 percent of all marine life remains in a pristine state.

In 1950, global annual fish catches were approximately 20 million tons; by 1970, this number had risen to 65 million tons, and, by 2005, it was 90 million tons. Larger scale fishing operations with bigger boats and nets and better technology have caused noteworthy collapses of fishery stocks in various regions, including Atlantic cod (*Gadus morhua*) in 1992, Atlantic swordfish (*Xiphias gladius*) in 1999, Pacific salmon (*Oncorhynchus sp.*) in 2001, and Mediterranean and Indian tuna (*Thunnus thynnus*) in 2007. More nets and lines in the water also have resulted in the deaths of approximately 300,000 sea turtles and sea mammals, such as seals and whales, each year.

Two 35-ton fin whales (*Balaenoptera physalus*) are bound by their tails to a Danish boat after being caught off the coast of Iceland. In 2009, Denmark officially requested permission, on behalf of subsistence communities living in Greenland, to allow Greenlanders to resume hunting ten humpback whales per year. As of early 2010, this item was under consideration by the International Whaling Commission, the regulatory body that oversees the hunting of whales. (*AFP/Stringer/Getty Images*)

In 2008, nearly half the humans on Earth, upward of 2.6 billion people, relied on ocean fish for their primary protein source. An expansion of commercial fishing has benefited communities and helped feed them. Fishing boats have grown significantly larger. Larger fleets (approximately 4 million vessels worldwide) have employed more people. Improved fishing technology includes bigger, more efficient nets and longline systems capable of hauling thousands of large fish in a single catch.

Worldwide demand for fish has resulted in the use of boats able to generate power for icemakers and refrigeration, allowing fishermen to spend months at sea, and such technologies as advanced satellite communications to pinpoint fishing grounds and sonar to locate schools of fish. The worldwide commercial fishing industry employs 210 million people and earns nearly $200 billion annually.

Despite, or perhaps because of, these advances, the negative impacts of large-scale fishing have been widespread. In 1980, 1,736 species were sought commercially, and up to 14 percent of world fisheries had reached some stage of collapse. In 2006, up to 7,784 species were caught for sale, and approximately 29 percent of fisheries were in a state of collapse. With declining catches close to shore, commercial fishing gradually turned to deeper waters, threatening additional species and further affecting the marine food chain.

It takes approximately 100,000 pounds (45,359 kilograms) of plankton to feed 10,000 pounds (4,536 kilograms) of small marine life, which in turn nourishes 1,000 pounds (454 kilograms) of small fish, which feeds 100 pounds (45 kilograms) of medium fish, which produces 1 pound (0.45 kilograms) of large fish that humans consume. Overfishing can sever a link anywhere in this chain, resulting in fewer species overall.

One of the more stark examples of the effects of overfishing can be seen in the demise of the whale, a sea mammal that has roamed the oceans for approximately 54 million years. Hunting and capturing an 85-foot (26-meter), 50,000-pound (22,680-kilogram) whale provided food and oil for months for a small coastal community (such as Iceland or Alaska) beginning in the eighteenth century. In the mid-nineteenth century, sailing ships caught approximately 3,000 whales in a slow process of hand harpooning and being taken for a "sleigh ride" by rowboats until the whale tired and could be slaughtered. In the twentieth century, boat engines, first powered by steam (fed by coal) and then by oil, were able to capture multiple whales in a day.

In 1910, approximately 13,500 whales—mostly North Atlantic right whales (*Eubalaena glacialis*), blue whales (*Balaenoptera musculus*), fin whales (*Balaenoptera physalus*), and sperm whales (*Physeter catodon*)—were caught globally for food; lamp oil was made from the blubber and manufactured products (such as combs, jewelry, and other goods) from the bones and other parts of the whale. In 1938, nearly 53,000 whales were captured, and whaling fleets added minke whales (*Balaenoptera acutorostrata*), gray whales (*Eschrichtius robustus*), bowhead whales (*Balaena mysticetus*), and humpback whales (*Megaptera novaeangliae*) to the long list of sought-after species.

In 1946, the International Whaling Commission (IWC) was created to maintain fishery stocks and help develop the industry. The need for cheaper food and a source of fat for margarine brought a steady climb in whaling. In 1963, up to 63,000 whales were hunted commercially. By 1972, whale populations had decreased substantially, and species such as the gray whale and bowhead whale had become critically endangered.

By the early 1970s, the IWC's membership had grown to include a number of nations that opposed commercial whaling (effective in 1986). In 1982, the commission voted, twenty-five to seven, to ban whaling. Only the industrial nations of Iceland, Japan, and Norway continue to hunt whales on a large scale, and only smaller aboriginal settlements (such as those in Asia, the Caribbean, North America, and Russia) still hunt them for subsistence.

The 28-year-old conservation policy has been a success; despite the impacts of boats colliding with whales, reduced food populations, and lowered water quality, some whale populations are increasing, albeit slowly. The population of the blue whale, which has been estimated at 275,000 in the 1700s, plummeted to around 2,000 by 1965, but it had increased to approximately 5,000 by 2009.

Another species, the North Atlantic right whale, has yet to show a stable recovery, and the critically endangered population of fewer than 400 whales may not be able to survive. Threats also loom for this species from the effects of climate change. Warming ocean temperatures and a reduction of plankton food stocks hinder the full recovery of the whale population to pre-1700 levels. In 2010, IWC members reached an advanced stage of negotiating a new whale ban that would, after a period of ten years, halt whale hunting worldwide—even by Iceland, Norway, and Japan—but would allow subsistence hunting to continue.

Oil Extraction

The discovery of oil in the marine environment and the resultant offshore drilling have created considerable wealth for nations such as Brazil, the Democratic Republic of the Congo, Great Britain, Norway, Russia, and Venezuela. The global economy consumed up to 4.6 billion gallons (17.4 billion liters) of oil per day in 2008; up to 30 percent of that oil is extracted from underwater sources.

In the early twentieth century, most ocean wells were dug in shallow waters less than 1,000 feet (304 meters) deep. As fewer new wells were discovered in shallow waters, oil operations pushed into deeper waters. In the Gulf of Mexico, thirty-nine deepwater oil fields pump up to 52 percent of the oil produced. The

ENDANGERED WHALES

Name	Scientific Name	Size (Length, Weight)	Status
Blue Whale	Balaenoptera musculus	110 feet, 181 tons	Between 8,000 and 14,000.
Fin Whale	Balaenoptera physalus	66 feet, 75 tons	40,000 counted in the North Atlantic; and fewer than 3,000 in the Southern Hemisphere.
North Pacific Right Whale	Eubalaena japonica	60 feet, 100 tons	Between 100 and 300 whales counted, almost extinct.
North Atlantic Right Whale	Eubalaena glacialis	60 feet, 100 tons	400 to 450 whales counted, nearly extinct, population falling.
Sei Whale	Balaenoptera borealis	66 feet, 45 tons	54,000, mostly in deep-ocean nonpolar or tropical waters.

Source: The World Conservation Union, 2007.

"Jack Number 2" well, for example, is located 175 miles (282 kilometers) offshore in 7,000 feet (2,134 meters) of water and extracts oil from 28,175 feet (8,588 meters) down. It provides 330,000 gallons (1,249,186 liters) of oil per day.

Large metal "traps" are placed at the ocean floor to limit drilled oil from pouring out into the ocean. Other devices separate water from oil to reduce contamination during pumping operations. Despite these and other safeguards, releases from oil drilling platforms have caused considerable harm to the oceans. Drilling fluids, lubricants, and cutting solvents contain high levels of arsenic, aluminum, cadmium, copper, lead, and other heavy metals. This toxic soup can grow to form an oily blob around a drilling rig, eventually reaching about 3 feet (0.9 meter) in thickness and 100 feet (30 meters) in diameter. If oil rig staff do not capture this oil, ocean-bottom currents slowly carry off the material, impacting marine life. Worldwide oil released from offshore rigs is estimated at 15 million gallons (57 million liters) per year.

Oil Spills

Of the approximately 706 million gallons (2,673 million liters) of waste oil released per year in the World Ocean, land-based pollution (from factories and storm water runoff) is responsible for nearly 51 percent, offshore drilling operations and tanker spills contribute approximately 8 percent, natural oil

OIL POLLUTION IN OCEANS

Type of Release	Amount Each Year (Gallons)
Natural bedrock releases (underwater)	47 million
During ocean oil removals	880,000
During ocean oil transport	2.7 million
Pollution from upland sources (such as factory releases, street system storm water outflow, spills)	25 million
Total:	75.58 million

Source: *Oil in the Sea*, National Research Council, Washington, DC: National Academies Press, 2003.

releases are estimated at 8 percent, deposition of hydrocarbon pollution from airborne sources equals 13 percent, and ship maintenance and releases account for 20 percent.

Large ocean oil spills began occurring in the 1970s and have been a prevalent problem ever since. The largest accidental oil spill in history occurred in the Persian Gulf in 1991. Approximately 240 million gallons (908 million liters) of oil spilled into the ocean near Kuwait and Saudi Arabia when several tankers, port facilities, and stor-age tanks were destroyed during Gulf War operations. The second-largest accidental oil spill, the blowout of the Ixtoc I exploratory well off the shores of Mexico in 1979, gushed 140 million gallons (530 million liters) of oil into the Gulf of Mexico.

By comparison, the *Exxon Valdez* tanker spill in 1989 poured 10.8 million gallons (40.8 million liters) of oil into Prince William Sound, Alaska. Although the *Exxon Valdez* accident did not release as much oil as a fixed platform can, it still did a substantial amount of damage to the biotically rich local environment, received a large amount of publicity, and raised worldwide awareness of oil spills. Reforms since the *Exxon Valdez* incident include the United Nations require-ment, effective in 2005, that all oil tankers must have double-walled tanks to protect against spills if a ship should run aground.

An underwater oil well blowout on April 20, 2010 at the Deepwater Horizon offshore drilling platform located 40 miles (64 kilometers) off Louisiana in the Gulf of Mexico proved the dangers of drilling in 5,000 feet (1,524 meters) of water. The facility, which was owned by BP, was demolished in an explosion that claimed the lives of eleven rig workers.

In addition to the oil leaking onto the water's surface, scientific researchers found several very large underwater plumes. One measuring 10 miles (16 kilometers) long by 3 miles (4.8 kilometers) wide, by at least 300 feet (91.4 meters) thick, sat 2,300 feet (701 meters) below the surface. The ruptures eventually were plugged, sealing the wellhead site.

The BP spill, which spread out over at least 2,500 square miles (6,500 square kilometers), resulted in a short-term fishing ban covering up to one quarter of all gulf waters. While it may take years for experts to determine the total volume of oil released, the spill has been recognized as the largest such incident in U.S. history, and it is certain to exceed tens of millions of gallons. Economic damage to the tourism and fishing industry may exceed $5 billion. Federal officials, facing public outrage due to lax enforcement of oil operations, planned a reorganization of the agency that regulates offshore oil and gas operations.

Pollution from Land Sources

Humans have dumped up to 80 percent of all ocean pollution from their upland trash and road sources. A city of 5 million residents (about the size of Miami, Florida) pollutes enough oily waste through street runoff and land-filling each year to equal the capacity of one large oil tanker.

Pollution includes toxic industrial waste, domestic garbage, medical waste, agricultural runoff, radioactive materials, storm water runoff, and human waste. Most pollution undergoes little biological or chemical breakdown while preserved in an underwater environment, due to the lack of microbes and ultraviolet light.

Excess nutrients such as nitrogen and phosphorus cause algae blooms that lead to depleted oxygen in water, killing fish and marine life. Garbage mixes with sediment and layers the bottom of the ocean, creating a hazardous habitat. Toxic waste, including copper, lead, mercury, and oil, enters and poisons the bodies of fish and other marine life.

Lack of Knowledge

Two historical preconceptions have played a central role in the degradation and overuse of marine resources. The first is the idea that the oceans are a "free" commodity universally harvestable by nations without limitations; such a philosophy makes regulatory practices difficult to create and enforce.

The second issue relates to how the oceans are viewed from land. It is difficult to see under the surface of the water, especially into the darker depths. Without straightforward visual images, oceans represent the unknown. Because of lack of exposure and understanding, humankind often has ignored or feared the ocean environment. Since the mid-twentieth century, for example, governments have expended more resources to map the surfaces of Mars and the moon than the world's oceans.

The development of maps that feature centrally placed oceans with detailed features are contributing to a changing mind-set. Multiple government space agencies have used satellite instruments to study the oceans from space and to establish baseline environmental data, against which scientists can compare future observations and monitor the health of this globally vital resource.

Selected Web Sites

Food and Agriculture Organization of the United Nations, The State of World Fisheries and Aquaculture: http://www.fao.org/fishery/sofia/en.

International Hydrographic Organization: http://www.iho-ohi.net/english/home/.

International Panel on Climate Change: http://www.ipcc.ch/.

International Union for Conservation of Nature, Red List: http://www.iucn redlist.org.

International Whaling Commission: http://www.iwcoffice.org.

National Geophysical Data Center, Marine Geology and Geophysics: http://www.ngdc.noaa.gov/mgg/image/2minrelief.html.

National Ocean and Atmospheric Administration, Ocean Dumping: http://oceanexplorer.noaa.gov/explorations/deepeast01/background/dumping/dumping.html.

Further Reading

Allsopp, Michelle, et al. *State of the World's Oceans*. New York: Springer, 2009.

Chivian, Eric, and Aaron Bernstein, eds. *Sustaining Life: How Human Health Depends on Biodiversity*. New York: Oxford University Press, 2008.

Estes, James A., et al., eds. *Whales, Whaling, and Ocean Ecosystems*. Berkeley: University of California Press, 2007.

Patton, Kimberley. *The Sea Can Wash Away All Evils: Modern Marine Pollution and the Ancient Cathartic Ocean*. New York: Columbia University Press, 2006.

Siedler, Gerold, et al., eds. *Ocean Circulation and Climate: Observing and Modelling the Global Ocean*. San Diego, CA: Academic Press, 2001.

United Nations Food and Agriculture Organization. *State of World Fisheries and Aquaculture 2008*. Rome, Italy, 2008.

World Conservation Union (International Union for Conservation of Nature). *The IUCN Red List of Endangered Species*. Gland, Switzerland, World Conservation Union (IUCN), 2009.

Worm, Borris, et al. "Impacts of Biodiversity Loss on Ocean Ecosystem Services." *Science* 314 (November 2006): 787–790.

12 Future of the Oceans

Today's oceans are relatively young geologic formations that are constantly undergoing changes in their physical size and dimension. Radiocarbon scans of submerged rock beds found that almost all ocean bottom is less than 180 million years old. While new oceanic bedrock is formed along oceanic ridges, it is simultaneously crushed in subduction zones and remelted into magma.

Scans also have revealed that planetary tectonic plates are moving, and fifteen major tectonic plates (and up to forty-one smaller ones) fit together like a temporary jigsaw puzzle. Sliding on a molten layer below the lithosphere, both the oceanic and continental plates collide with one another continually.

As the Atlantic Ocean grows, the Pacific Ocean shrinks. Within a few hundred million years, the surface of the planet will look considerably different than it does today. This oceanic geological sequence has been repeating itself for at least 3 billion years—and scientists predict it will continue to do so until the interior of the planet cools into a solid mass.

Other physical changes have happened to the oceans within a shorter time frame. Sediment and organic debris have accumulated and added layers to ocean and sea bottoms. In some places, this amounts to 15,000 feet (4,572 meters) of deposits, as in the Mediterranean Sea. The World Ocean has steadily worn away at the continental upland, carrying rock, sand, and minerals into the water.

As the polar ice caps have grown and shrunk according to shifts in surface temperatures, ocean water levels have fluctuated. During the Earth's ice ages, the oceans became smaller, revealing features such as the Siberian Land Bridge between Russia and North America. Global warming historically has meant a reduction in polar ice and widespread flooding across lowland regions, leaving places such as the Florida peninsula underwater for tens of thousands of years.

More recent changes such as human-induced global warming have become threats to the current stability in the global ocean environment. Two of the biggest changes that are impacting the oceans today are melting polar ice from higher air temperatures and ocean acidification from too much carbon dioxide pollution; both have been directly correlated with the substantial increase in greenhouse gas emissions since 1850.

In 2007, the International Panel on Climate Change reported that the temperature increases could melt the entire northern ice cap in the Arctic Sea by the summer of 2040, and winter ice depth may shrink drastically. The panel also concluded that by the year 2100, carbon dioxide emissions likely will cause ocean pH levels to decrease by as much as 0.5 pH units—the lowest they have been in the last 20 million years—threatening tens of thousands of marine organisms such as corals, crabs, and oysters, which rely on available calcium to form shells or exoskeletons.

Scientists theorize that rising temperatures may cause oceans to expand away from the equator and toward the poles. Most of the expansion will take place in the North Atlantic Ocean, near the North Pole. Because the poles are closer to the Earth's

Rescuers use a crane, fitted with giant slings, in an effort to transport this long-finned pilot whale (*Globicephala melas*) back out to sea from Hamelin Bay, which lies south of the city of Perth, Australia. This whale and ten others survived a mass beaching on Australia's west coast that claimed ninety-nine long-finned pilot whales and ten bottlenose dolphins (*Tursiops truncates*). The reason for this 2003 and other such events is not known, although some scientists suggest a widespread bacterial infection or excessive underwater sonar use by military operations as possible causes. *(Tony Ashby/Stringer/AFP/Getty Images)*

axis of rotation, having more mass in these locations could speed up the planet's rotation.

Scientists theorize that at least 1 million species that inhabit ocean waters will continue to change in population as they either adapt or fail to adapt to a complex set of pressures such as weather fluctuations, food shortages, chemistry changes, and temperature elevations.

Protecting the Ocean Environment

Governments, nongovernmental organizations and international groups are making a concerted effort to protect the world's oceans. A "World Ocean Day" was proposed by Canada at the United Nations Earth Summit in Rio de Janeiro in 1992 in order to coordinate educational and advocacy events on a global scale. In 2005, Oceana—the largest international organization focused solely on ocean conservation—reached agreements with commercial and recreational fishermen, federal fishing management council members, and the U.S. National Oceanic and Atmospheric Administration (NOAA)'s National Marine Fisheries Service staff to protect deep-sea coral and sponge communities in North America from destructive trawling and dredging fishing gear.

In November 2008, Healthy Reefs for Healthy People, an international, multi-institutional effort tracking the health of the Mesoamerican Barrier Reef, released an Eco-Health Report Card for the reef—its first comprehensive health assessment. The reef stretches from the northeast tip of Mexico's Yucatán Peninsula to Amatique Bay, Guatemala, making it the second-largest coral reef system in the world after the Great Barrier Reef in Australia. The Eco-Health Report Card found that overdevelopment in many communities had caused almost 48 percent of the reef system to be in a poor state and 6 percent of the system to be in a critical state where species loss is imminent.

In 2009, the National Marine Fisheries Service instituted federal regulations that banned large-scale fishing for krill in U.S. Pacific waters in order to prevent a catastrophic gap in the food chain through overfishing. Also in 2009, the United Nations General Assembly, after seventeen years of informal annual celebrations, designated June 8 as World Oceans Day. A weekend before that day's first official observance, the governors of the states of Delaware, Maryland, New Jersey, New York, and Virginia signed an agreement committing the states to a cooperative effort to protect the ocean waters of the mid-Atlantic.

Around the world, such groups and organizations are working independently and together to protect corals, stop illegal oil dumping, halt destructive trawling, save endangered species such as sea turtles and whales, and advocate for improved management and protection of the ocean for its long-term health.

Ocean Research Facilities

In 1950, there were a dozen scientific ocean research facilities, almost all of which were associated with government agencies or universities. By 2009, the number of research facilities had increased to thirty-two large centers spread across Africa, the Americas, Antarctica, Asia, and Europe. Their accomplishments include studying the ocean's influence and ocean phenomena, mapping ocean bottoms, analyzing water, inventorying thousands of species, and learning the fundamentals of this vast environmental biome.

Researchers also have gained a greater understanding of how human activities impact the oceans in a negative way. Scientists and educators have shared this information with government and public officials, resulting in new policies and laws intended to protect oceans and their biota.

Ocean Preserves

Although the international community has protected up to 12 percent of combined upland (forests, fields, wetlands, and mountains) from development, it has not done the same for ocean territory. Less than 1 percent of worldwide saltwater habitats have been designated as protected. This statistic is slowly changing, however, with an increase in global awareness of the importance of ocean habitat. The end of the twentieth century and beginning of the twenty-first century have seen a push to set aside larger areas of the world's oceans as preserves, where human development may be heavily regulated or even entirely banned.

Two noteworthy examples of efforts to create and manage ocean preserves include those in the Caribbean and the United States. In 1995, the Saint Lucian government created the Soufriere Marine Management Area, located off the island of Saint Lucia in the Caribbean Sea, to restore overfished tropical fisheries. The preservation area, which spans 6.8 miles (11 kilometers) of coastline was carved into zones with five different levels of protection. Within a closed reserve, after only three years' time, some fishing areas were found to experience a threefold increase in biomass.

The United States took a bold step in 2006 when it created a Pacific Ocean sanctuary called the Papahanaumokuakea Marine National Monument that spans 131,250 square miles (339,938 square kilometers) of ocean waters, islands, and atolls in the Hawaiian Islands. This area featured stronger federal protections to enhance conservation efforts for its endangered monk seals, nesting green sea turtles, and some 7,000 other species. All recreational fishing is banned within the monument, and, by 2011, all commercial fishing in the area will end, marking a five-year phaseout period.

Species Inventory

Although numerous scientific studies have been conducted to catalog up to 215,000 species from the oceans over the last 200 years, no modern inventory was undertaken until 2000. The Census of Marine Life is a network of 2,000 international scientists from eighty nations committed to a ten-year initiative to capture the diversity, distribution, and abundance of marine life in the oceans. The project plans to describe and catalog up to 14 million species, including those that might be extinct but can be sampled from bottom debris. In its first few years, the census added some 5,600 new species while completing fourteen field projects to assess major habitat areas. These areas included abyssal plains, continental margins, coral reefs, gulf regions, shore areas, and other ocean habitats. The first reports of seventeen projects spanning six ocean locations will be published in late 2010.

Renewable Energy

The majority of global electricity comes from five sources: coal, hydropower, nuclear fission, oil, and wood. Nearly all rely on land-based sources and generate significant pollution or otherwise impact the environment. The oceans, however, offer a promising source for considerable renewable energy with far

The Seagen tidal power generator is shown here with its turbine blades out of the water at the narrows of Strangford Lough in Northern Ireland. Marine Current Turbines, the company that created the alternative-energy power plant, eventually hopes to achieve large commercial production. *(Bloomberg/Getty Images)*

less pollution, centered around abundant opportunities for wind, current, and wave energy. Twenty-eight wind farms, all with at least a 2-megawatt capacity, have been built in offshore areas around the world, with a total capacity of 1,684 megawatts of power (enough to support 1.3 million homes), another seventeen are under construction, and forty-six projects currently are in advanced permitting stages.

The largest such installation is the Horn's Rev built 18.6 miles (30 kilometers) off the coast of Denmark in the North Sea in 2009. This 209-megawatt facility contains ninety-one turbines and generates enough power to support 200,000 homes. Denmark has developed 619 megawatts of wind-generating capacity across eight large offshore farms. Wind provides almost 20 percent of the country's power. In 2009, the Danish government began a project to grow this system by 50 percent by 2025 to reach a goal of providing 70 percent of the nation's power through wind.

Oceans also offer sources other than wind for power. The Rance Estuary in northern France has harnessed tidal flow since 1966, providing a daily average of 68 megawatts of power through a series of hydro-turbines. This system currently provides less than 1 percent of France's power.

The SeaGen turbine in Strangford Lough in Northern Ireland was built in a North Sea location that has several narrow channels between open water and fast-moving currents that exceed 13.1 feet per second (4 meters per second). The project, which was completed in 2007, produces 1.2 megawatts of power, relying on submerged rotors that resemble a windmill placed underwater.

In 2000, new technology using wave energy was put to use on the island of Islay, off Scotland's western coast. This system uses waves to turn turbines, generating a half megawatt of power. Each Islay turbine looks like a large floating buoy. The center rides up and down on the passing waves, and the motion causes air pressure to turn a turbine, which generates electricity. The first large-scale commercial project using this technology, the Agucadoura Wave Farm, was built in 2006; it is located in the Atlantic Ocean, 3 miles (5 kilometers) off the shore of northern Portugal near the city of Povoa de Varzim. Three energy converters can generate 2.2 megawatts of power. A total of thirty-one converters are planned to be installed in the area in 2010.

The Future of Ocean Resource Management

The collapse of an ocean fishery has occurred several times around the world since the 1980s. It begins with reduced catch levels. Regions or countries often end up in conflict over dwindling fish stocks. A system of quotas may be agreed upon by industry or government leaders to better manage the population.

Sometimes the quota system is imposed too late, and because reproductive pools already are too small, fish stocks barely improve or continue to shrink in popluation. Occasionally, the population challenge of one species (such as Atlantic striped bass, *Morone saxatilis*, in the North Atlantic) is related to the overfishing of another species (such as Atlantic herring, *Clupea harengus*) in the same area.

As a result, governments have turned to other methods of managing their ocean resources, including strict fish lengths and quotas, closing waters, buying out fishing fleets, financially supporting the aquaculture (aquafarming) industry, and passing other laws to protect entire ecological regions.

Catch Limits and Fishing Closures

Ocean managers and government officials have learned that nothing is as beneficial as restricting catch amounts or stringent fishing closures to help ecosystems naturally repopulate. This perspective has been applied to multiple areas, including waters between Canada and the United States in the North Atlantic (for cod) in 1992, waters between New Zealand and Australia (for orange roughy) in 2007, and in European Union waters (for bluefin tuna) in 2007.

Unfortunately, the impact of a sudden closure can be devastating economically. The fishing fleets may be able to change their gear and pursue different species, but in many cases, they are simply forced out of business.

Government subsidies of fishing fleets have helped them through difficult economic times, but such subsidies also have been shown to contribute to bloated fishing fleets and exploitive fishing practices. A 2009 study of the U.S. fisheries, *Quantification of U.S. Marine Fisheries Subsidies*, found that annual subsidies exceeded $713 million and could be worth as much as a fifth of the value of the fish caught. The study also concluded that some 56 percent of subsidies to fishermen actually lead to overfishing, by lowering overhead costs such as fuel prices (by eliminating federal and state taxes). The western United States received particular attention in the report, as the region catches only 2 percent of the total U.S. catch, but receives up to 23 percent of the total subsidies.

Governments sometimes initiate buyouts as an incentive to change the industry. The U.S. Congress authorized several in 1996. One buyout included $46 million to purchase ninety-two of the 260 ground-fishing boats in the northern Pacific, off the states of Alaska and Washington, to protect flounder, snapper, and whiting stocks.

Aquaculture

The farming of freshwater and saltwater organisms (including fish, mollusks, crustaceans, and aquatic plants), a practice known as aquaculture, is performed in large coastal pens, indoor pools, or outdoor ponds. Over the last half century, the industry has grown as a whole, particularly in the amount of fish that is farmed due to reduced fishing in the open oceans.

In 1950, less than 1 million tons of farmed fish were provided globally; since 1985, the industry has grown steadily by nearly 9 percent a year. By 2003, in Japan alone, total aquaculture production was estimated at 1.3 million tons, and cultivation worldwide totaled 42 million tons. As of 2010, aquaculture accounted for approximately a third of the total fish harvest.

Fish farming has not come without its own challenges, including complex water quality and disease management issues. Environmental impacts have included wetlands removals, harmful algae blooms from waste streams, and loss of "wild" qualities in certain fish. For example, farmed salmon meat often loses the bright pink color typical of wild salmon.

Consumer Education

Another approach to better manage ocean fish stocks has been to engage the consumer in making smart choices about buying sustainable seafood. The Marine Stewardship Council, a nonprofit organization created in London, England, in 1997, was formed "to safeguard the world's seafood supply by promoting the best environmental choice."

By 2009, the council had given thirty-eight fisheries around the world the certified designation of "sustainable." Each certified location has to undergo a complex review (every one to two years), proving that fish are being removed at a rate that is sustainable.

One example, in the stormy Southern Ocean, includes the harvesting of mackerel icefish (*Champsocephalus gunnari*). Certified by the council in 2006, this cold-water species is caught about 1,500 miles (2,414 kilometers) west of Australia by boats that trawl with nets. These mackerel grow in three years to their maximum size of about 10 inches (25 centimeters) long. Under the management of the Australian government, only 220 tons of caught mackerel icefish were permitted in 2008. (The government changes this number every year, raising it to as high as 2,980 tons in 2002 or dropping it to as low as 42 tons in 2006.) Each area certified by the council is reinspected at least annually to make sure that it is in compliance with agreed-upon management goals and catch limits.

Treaties and Laws

The overarching international treaty that provides a structure of law and procedures for oceans is the United Nations Convention on the Law of the Sea (UNCLOS, also called the Law of the Sea Convention or the Law of the Sea

Treaty). The first version was drafted in 1958, and it was updated substantially in 1982 and 1995. The original convention laid out a sovereign territory that extended up to 12 miles (19 kilometers) off the coast of any ocean-bordering country.

In 1958, the United Nations also created a Convention on the High Seas, an international treaty created to codify the rules of international law and address ocean procedures outside of territorial waters. High seas (defined as water bodies outside of any territorial or jurisdictional claim) were open to all countries for navigation, fishing, and economic development efforts such as laying cables or drilling for oil.

Due to competition among nations since the mid-twentieth century, in 1982, the United Nations expanded the sovereign limit of jurisdiction to a 200-mile (322-kilometer) Exclusive Economic Zone. This revision also established a number of standards to protect the marine environment, allowed scientific research throughout the world's oceans, and addressed the issue of deepwater mineral drilling by creating a new International Seabed Authority to regulate such practices.

Although UNCLOS is a wide-reaching treaty that brought organization and standards to the international community, it did not erase disputes between countries in specific regions. Although some arguments centered over island ownership, most were over fishing grounds and concerned the expansion of sovereign territory.

The 1950s and 1970s saw cod wars between Great Britain and Iceland. The 1980s brought tuna conflicts between France and Great Britain. In 1992, a dispute between Morocco and the European Union resulted in Morocco imposing fishing tariffs and limits on European fishing trawlers in that nation's territorial waters. Several European countries, citing historical ownership of parts of Morocco, felt that the government's claim of a 200-mile (322-kilometer) exclusive economic zone was unjust. In 1995, a Canadian government boat seized a large Spanish trawler for illegally fishing in its coastal waters. As a result of such actions, many fishing fleets necessarily found themselves seeking their catch in deeper waters.

UNCLOS was amended in 1982 and again in 1995 with a series of regulatory measures titled the Conservation and Management of Straddling Fish Stocks and Highly Migratory Fish. These successive modifications set new international standards to reduce both coastal and deep-sea catches. Even though most countries have imposed fishery management plans within their 200-mile (322-kilometer) economic zones to protect stocks, unregulated deep-ocean fishing is common. Enforcement is very difficult, even with satellite imaging. These deepwater ocean harvests pose the next challenge for long-term fisheries conservation and the health of ocean ecosystems.

Selected Web Sites

Census of Marine Life: http://www.coml.org.
International Fish Database: http://www.fishbase.org.
International Ocean Institute: http://www.ioinst.org.
International Seabed Authority: http://www.isa.org.jm/en/home.
United Nations Oceans and Law of the Sea: http://www.un.org/Depts/los/.

Further Reading

Christie, Donna R., and Richard G. Hildreth. *Coastal and Ocean Management Law in a Nutshell.* St. Paul, MN: Thompson West, 2007.

Cramer, Deborah. *Smithsonian Ocean: Our Water, Our World.* Washington, DC: Smithsonian Books, 2008.

Crist, Darlene Trew, Gail Scowcraft, and James M. Harding, Jr.. *World Ocean Census: A Global Survey of Marine Life.* Buffalo, NY: Firefly, 2009.

Earle, Sylvia A., and Linda K. Glover. *Ocean: An Illustrated Atlas.* Washington, DC: National Geographic Society, 2009.

Ellis, Richard. *The Empty Ocean: Plundering the World's Marine Life.* Washington, DC: Island, 2003.

Erickson, Jon. *Marine Geology: Exploring the New Frontiers of the Ocean.* New York: Facts on File, 2002.

McLeod, Karen, and Heather Leslie, eds. *Ecosystem-Based Management for the Oceans.* Washington, DC: Island, 2009.

Norse, Elliott, and Larry Crowder. *Marine Conservation Biology: The Science of Maintaining the Sea's Biodiversity.* Washington, DC: Island, 2005.

Roberts, Callum. *The Unnatural History of the Sea.* Washington, DC: Island, 2007.

Sharp, Renee, and Rashid Sumaila. "Quantification of U.S. Marine Fisheries Subsidies." *North American Journal of Fisheries Management* 29 (2009): 18–32.

Abiogenesis. The study of how life on Earth formed from inanimate matter. One such theory—the iron-sulfur world theory—posits that life formed during a succession of events at the ocean bottom along a volcanic rift.

Abyss. The bottom of an ocean where the depth is greater than 6,000 feet (1,755 meters).

Abyssal plain. The mostly flat region found at the ocean bottom between the mid-ocean ridge and the lower edge of the continental shelf.

Algae. A diverse group of photosynthetic, single-cell, and multicellular aquatic organisms. Most algae live at or near the surface of wetlands, freshwater bodies, or oceans, and are consumed by species such as copepods. There are approximately 15,000 different species of algae worldwide.

Anaerobic. Organisms that are able to live with little or no oxygen.

Aquifer. An underground layer of permeable gravel, rock, sand, or soil from which groundwater can be extracted.

Arctic Circle. A circle of latitude that runs 66°33' north of the equator, marking the edge of the Arctic Sea. The equivalent circle in the Southern Hemisphere is called the Antarctic Circle.

Atmosphere. The mixture of gases surrounding the Earth. By volume, it consists of about 78.08 percent nitrogen, 20.95 percent oxygen, 0.93 percent argon, 0.036 percent carbon dioxide, and trace amounts of other gases, including helium, hydrogen, methane, and ozone.

Atoll. An island built from a collapsed volcanic mountain (most underwater), surrounded by a coral reef, often in the shape of a circle.

Bacteria. Simple, single-celled microorganisms that live in organic matter, soil, water, and the bodies of plants and animals. Some bacteria are pathogenic—that is, responsible for causing infectious diseases.

Bathymetry. The measurement of an ocean's depth to determine its contours and geography.

Bay. A body of water that is formed by an indent in the shore between two capes or headlands. Typically, a bay is bigger than a cove but not as large as a gulf.

Beach. The landform where ocean and land meet. Characterized by a steep upward slope and loose, pulverized sand and small rocks, it is the zone of heaviest impact during coastal storms.

Benthic zone. The lowest level of a body of water, inhabited by animals or plants that have adapted to the environment and are able to tolerate low temperatures, a lack of sunlight, and low oxygen levels. The shallowest benthic areas are around 600 feet (183 meters), and the deepest are more than 19,600 feet (5,974 meters).

Biodiversity. Biological diversity, indicated by the variety of living organisms and ecosystems in which they live.

Biome. A regional ecosystem that is characterized by a particular climate and biological community.

Biosphere. The ecological system made up of all living organisms on Earth.

Black smokers. Geologic volcanic formations along mid-ocean ridges where water emerges from the bedrock in high-velocity jets. The hot emissions, which may exceed 662 degrees Fahrenheit (350 degrees Celsius), often are dark in color and laden with metals and sulfides.

Brackish. A mixture of freshwater and salt water.

Carbon dioxide (CO_2). A gas that is found naturally in the atmosphere at a concentration of 0.036 percent. It is fundamental to life and is required for respiration and photosynthesis.

Carbon neutral. A state at which no carbon is released into the environment, achieved through a process of controlling the emission and absorption of carbon.

Carbon sink. A natural or man-made reservoir or habitat (such as forests, oceans, and wetlands) that can store carbon for long periods of time.

Cephalopod. An animal that lives in benthic zones, possesses a large head and eyes, and has a circle of arms (or tentacles) around a mouth (such as octopus and squid).

Cetacean. An aquatic mammal species that includes the family of dolphins, porpoises, and whales.

Climate change. Global variation in the Earth's climate over time. A recent phenomenon (measured over the last two centuries), it is attributable largely to human atmospheric pollution. Climate change results in temperature increases and other shifts in weather patterns, severe droughts and floods, rising sea levels, and biotic changes (often causing the loss of species in an area).

Conservation. The protection, preservation, management, or restoration of natural resources, plants, and wildlife.

Continental shelf. A terrace of land at the perimeter of a continent. Typically, it has a gentle pitch leading from the shallow edge of the continental plate to the deeper ocean plate.

Crustacean. A family of up to 55,000 aquatic species, including barnacles, crabs, lobsters, and shrimp. Most have a tough exoskeleton that can be shed to allow growth.

Current. The flow of a body of water in a defined size and direction, powered by three distinct factors: thermal influence from the sun, tides, and winds.

Decomposition. The process by which organic matter is broken down by dynamic forces such as weather, water, and microbes.

Diatoms. Microscopic members of the algae family and a common type of phytoplankton. These animal bodies are built from silica and shaped into two distinct valves (or shells). There are up to 100,000 species of diatoms, and they provide up to 45 percent of the primary production for the oceanic food web.

Dilution method. The dumping of waste into water bodies or groundwater wells, where it is diluted by the movement of water and, theoretically, made less potent.

Discharge. The release of untreated waste into waterways, often following excessive precipitation that exceeds the capacity of a sewer system or waste processing facility.

Ecosystem. The interaction of a biological community and its environment.

Eddy. A circular current of water that runs counter to the main current. Examples include the large North Pacific Ocean gyre and the smaller Sea of Japan eddy.

Endangered (and threatened) species. A species that is in danger of extinction because the total population is insufficient to reproduce enough offspring to ensure its survival. A threatened species exhibits declining or dangerously low population but still has enough members to maintain or increase numbers.

Endemic. A plant or animal that is native to and occupies a limited geographic area.

Environment. External conditions that affect an organism or biological community.

Erosion. The removal of materials from a location as a result of wind or water currents.

Estuary. A body of salt water located in a coastal area that has at least one freshwater body flowing into it. These biomes contain a diversity of freshwater and saltwater species.

Eutrophication. Algae blooms in aquatic ecosystems that are caused by excessive concentrations of organic and inorganic nutrients, including garden and lawn fertilizers and sewage system discharges. These blooms reduce dissolved oxygen in the water when dead plant material decomposes, causing other organisms, such as crabs and fish, to die.

Fathom. A unit of length used to measure ocean depth. A fathom is equal to 6 feet (1.83 meters).

Greenhouse effect. A warming of the Earth caused by the presence of heat-trapping gases in the atmosphere—primarily carbon dioxide, nitrous oxide, ozone, water vapor, and methane—mimicking the effect of a greenhouse.

Habitat. An area in which a particular plant or animal can live, grow, and reproduce naturally.

Hazardous waste. Industrial and household chemicals and other wastes that are highly toxic to humans and the environment.

High tide. The peak level of the daily flood tide.

Hurricane. A large tropical storm, also called a cyclone, that is formed over warm ocean waters, spinning in a counterclockwise direction.

Hydrologic cycle. The continuous movement of water as a result of evaporation, precipitation, and groundwater or surface water flows.

Hydrology. The science of the movement and properties of water both aboveground and underground.

Hydrophone. An electronic device that detects sound at varying frequencies, often used to identify underwater volcanic eruptions.

Hydrothermal vent. A location on the ocean bottom where superheated water emerges from cracks created by volcanic activity. Hot water—up to 400 degrees Fahrenheit (204 degrees Celsius)—pours from an opening, spilling mineral-laden deposits into the ocean.

Invertebrate. An organism without a spinal column. Invertebrates make up nearly 98 percent of all species (exceptions are mammals, birds, amphibians, reptiles, and fish). These include coral, sponges, arachnids, insects, jellyfish, octopuses, squid, starfish, snails, and worms.

Latitude. A geographic coordinate that indicates a position north or south of the equator, denoted by imaginary horizontal lines running in an east-to-west direction on maps. Latitude is an angular measurement expressed in degrees, ranging from 0° at the equator to 90° at each pole. Each degree is approximately 68 miles (109 kilometers) in distance.

Longitude. A geographic coordinate that indicates a position from east to west, denoted by imaginary vertical lines or meridians running in a north-to-south direction on maps. Longitude is an angular measurement expressed in degrees, where 0° is the Greenwich meridian line, located in London; measurements climb to 180° east or 180° west. Each degree is approximately 60 miles (97 kilometers) in distance.

Low tide. The lowest level of the daily flood tide.

Magma. A fluid and molten rock, ranging in temperature from 1,292 to 2,372 degrees Fahrenheit (700 to 1,300 degrees Celsius) and located beneath the Earth's crust, that is commonly ejected by volcanoes and other geologic structures.

Mid-ocean ridge. A large underwater mountain range that emerges from a divergent boundary between two tectonic plates. Pushed up from the Earth's mantle as hot magma, such ridges can be found along 40,000 miles (64,374 kilometers) of the ocean floor.

Mollusks. A large and diverse family of animals that live in the aquatic environment, including clams, octopuses, squid, and snails. Estimates of this species include up to 115,000 phyla.

Nautical mile. A measure of distance that is equal to one minute of latitude or 6,076 feet (1,852 meters).

Neritic. The shallow water region where an ocean meets the coast.

Non-point source. A source of pollution that comes from a broad area (such as car and truck emissions) rather than a single point of origin (known as a point source).

Nutrients. Chemicals that are necessary for organic growth and reproduction. For example, the primary plant nutrients are nitrogen, phosphorus, and potassium.

Ocean. The body of interconnected water that encircles the planet, broken down into five divisions: Atlantic, Pacific, Indian, Southern, and Arctic. Covering a total of 139 million square miles (360 million square kilometers), the World Ocean has an average salinity of 3.45 percent and an average depth of 12,430 feet (3,789 meters).

Oceanography. The study of the Earth's oceans and seas, including marine environments, organisms, dynamic systems, and geology.

Pelagic. The offshore ocean environment.

pH. A measurement of the acidity or alkalinity of a solution, referring to the potential (p) of the hydrogen (H) ion concentration. The pH scale ranges from 0 to 14; acidic substances have lower pH values, while alkaline or basic substances have higher values.

Photosynthesis. The chemical process by which carbon dioxide and water are converted into organic compounds, primarily sugars, when the chlorophyll-containing tissues of plants are exposed to sunlight. Oxygen is released as a by-product. Photosynthesis occurs in algae, phytoplankton, plants, and some bacteria.

Plankton. Aquatic microscopic plants and animals that live by floating at or near the surface of water in very large groups.

Plate tectonics. A theory that describes the large-scale movement of the lithosphere—the outermost part of the Earth's surface—which is made up of fifteen major and forty-one minor rock plates that fit together like a jigsaw puzzle. The formation, movement, and alteration of these plates in dynamic flux is the process of plate tectonics.

Point source. A source of pollution that originates from a single point, such as the discharge end of a pipe or a power plant.

Salt water. Freshwater that contains a percentage (0.5 parts per thousand) of dissolved salt minerals.

Sea. A body of an ocean, either enclosed or open, that possesses its own characteristics, such as water temperature, salinity, and species. There are currently sixty-six seas found in the five oceans.

Seamount. An underwater mountain built from oceanic volcanoes that usually does not reach the surface of the water. Nearly 30,000 separate peaks make up several dozen seamounts around the world. These steep, sloped formations host a diversity of corals and other marine life.

Sediment. Fine particles made of both mineral and organic ingredients that are deposited on the ocean bottom in layers.

Storm water runoff. The potent mix of dirt, nutrients, oils, and trash that is flushed from the land by rain, ending up in wetlands, lakes, rivers, streams, oceans, and groundwater.

Strait. A narrow channel of water between two sections of land that connects two larger bodies of water. (Examples include the Strait of Magellan and the Strait of Gibraltar.)

Subduction zone. An underground region where the collision of two or more tectonic plates creates volcanoes, earthquakes, and other geologic events. Typically, one section of tectonic plate is uplifted and the other is pressed downward into the subduction zone, where it is crushed and melts into underground magma.

Substrate. The layers on the bottom of a body of open water to which marine organisms anchor or attach themselves.

Succession. Changes in the structure and composition of an ecological community. Succession is driven by the immigration of new species and the competitive struggle among species for nutrients and space.

Taxa. The classification of organisms in an ordered, hierarchical system that indicates their natural relationships. The taxa system consists of the following categories: kingdom, phylum, class, order, family, genus, and species.

Tide. The predictable rise and fall of the level of water in a large body, resulting from powerful gravitational forces exerted by the moon, the sun, and other astronomical forces. Some regions have two tides a day, some have one large tide and one small tide, and others have only one tidal cycle.

Trench. A very deep depression found at the bottom of an ocean, formed from the subduction of an ocean plate where it meets a continental plate.

Tsunami. A large set of waves created by an underwater earthquake or volcanic eruption. A tsunami can stretch for hundreds of miles and travel up to 500 miles (805 kilometers) per hour, rising as high as 1,000 feet (305 meters) above the coast before breaking onto landforms.

Typhoon. A type of tropical storm system, similar in behavior to a hurricane, that is formed in the northwest Pacific Ocean.

Upwelling. A aquatic phenomenon in which wind pushes currents downward, lifting cold, dense, and nutrient-rich waters from lower levels to the surface. The types of upwellings include coastal, eddy, and deep ocean.

Water (H_2O). A tasteless, odorless liquid formed by two parts of hydrogen and one part of oxygen.

Water column. A vertical body of water running from the surface to the bottom of the ocean.

Wave. The movement of energy—from heat from the sun, wind in the atmosphere, or volcanic activity—into water, producing a physical disturbance. Waves can take shape in substantially different sizes, travel at slow or fast speeds, and last for less than a second or for days.

Wetland. A low-lying area where a high level of saturation by water provides a habitat that is critical to plant and animal life.

World Ocean. The interconnected system of the Earth's saltwater bodies, which holds an estimated volume of some 328 million cubic miles (1.4 billion cubic kilometers), or 343,423,668,428,484,681,262, or over 343 quintillion, gallons (1.3 sextillion liters, or the number 1.4 followed by 21 zeroes).

Zooplankton. Microscopic aquatic animals that float in open water and play a central role in the food chain. As invertebrates, they consume phytoplankton (plants) and are consumed by birds, fish, mussels, and mammals (such as whales).

Selected Web Sites

For additional Web Sites on more specific topics, please see the Web Site listings in individual chapters.

Alaska Department of Fish and Game: http://www.adfg.state.ak.us.

Alaska Marine Conservation Council: http://www.akmarine.org.

Alaska Oceans Program: http://www.alaskaoceans.net.

Alfred Wegener Institute for Polar and Marine Research: http://www.awi.de/en/home.

Australian Institute of Marine Science: http://www.aims.gov.au/.

British Oceanographic Data Centre: http://www.bodc.ac.uk.

Census of Marine Life: http://www.coml.org.

Food and Agriculture Organization of the United Nations, The State of World Fisheries and Aquaculture: http://www.fao.org/fishery/sofia/en.

Intergovernmental Oceanographic Commission: http://ioc-unesco.org.

International Council for the Exploration of the Sea: http://www.ices.dk.

International Hydrographic Organization: http://www.iho-ohi.net/english/home/.

International Ocean Institute: http://www.ioinst.org.

International Panel on Climate Change: http://www.ipcc.ch/.

International Seabed Authority: http://www.isa.org.jm/en/home.

International Whaling Commission: http://www.iwcoffice.org.

Mid-Atlantic Ridge Ecology Study (MAR-ECO): http://www.mar-eco.no.

National Geophysical Data Center: http://www.ngdc.noaa.gov/ngdc.html.

National Maritime Museum (United Kingdom): http://www.nmm.ac.uk.

National Ocean and Atmospheric Administration: http://www.noaa.gov/.

North Pacific Fishery Management Council: http://www.fakr.noaa.gov/npfmc.

North Sea Commission: http://www.northseacommission.info/.

The North Sea Region Programme: http://www.northsearegion.eu.

Ocean Research Institute, University of Tokyo: http://www.ori.u-tokyo.ac.jp/index_e.html.

Seas at Risk: http://www.seas-at-risk.org/.

Secretariat of the Antarctic Treaty: http://www.ats.aq/.

U.S. Geologic Survey: http://www.usgs.gov/.

The United Kingdom Offshore Oil and Gas Industry Association: http://www.oilandgas.org.uk.

United Nations Atlas of the Oceans: http://www.oceansatlas.org.

United Nations Educational, Scientific and Cultural Organization: http://www.unesco.org/.

United Nations International Maritime Organization: www.imo.org.

United Nations Oceans and Law of the Sea: http://www.un.org/Depts/los/index.htm.

Woods Hole Oceanographic Institution: http://www.whoi.edu.

Further Reading

Allsopp, Michelle, et al. *State of the World's Oceans*. New York: Springer, 2009.

Australian Government Mission and Maldives Marine Research Center. *Assessment of Damage to Maldivian Coral Reefs and Baitfish Populations from the Indian Ocean Tsunami*. Canberra, Australia: Commonwealth of Australia, 2005.

Ballesta, Laurent, and Pierre Descamp. *Planet Ocean: Voyage to the Heart of the Marine Realm*. Washington, DC: National Geographic Society, 2007.

Barale, Vittorio, and Martin Gade, eds. *Remote Sensing of the European Seas*. London: Springer, 2008.

Berkman, Paul Arthur. *Science into Policy: Global Lessons from Antarctica*. San Diego, CA: Academic Press, 2002.

Bindoff, Nathaniel, et al. *Observations: Oceanic Climate Change and Sea Level*. New York: Cambridge University Press, 2007.

Brand, Hanno, and Lee Muller, eds. *The Dynamics of Economic Culture in the North Sea and Baltic Region*. Hilversum, The Netherlands: Uitgeverij Verloren, 2007.

Buschmann, Rainer. *Oceans in World History*. New York: McGraw-Hill, 2006.

Byatt, Andrew, Alastair Fothergill, and Martha Holmes. *The Blue Planet*. New York: DK, 2001.

Chivian, Eric, and Aaron Bernstein, eds. *Sustaining Life: How Human Health Depends on Biodiversity*. New York: Oxford University Press, 2008.

Christie, Donna R., and Richard G. Hildreth. *Coastal and Ocean Management Law in a Nutshell*. St. Paul, MN: Thompson West, 2007.

Coleridge, Samuel Taylor. *The Rime of the Ancient Mariner*. 1798. Edison, NJ: Chartwell, 2008.

Cramer, Deborah. *Smithsonian Ocean: Our Water, Our World*. Washington, DC: Smithsonian Books, 2008.

Crist, Darlene Trew, Gail Scowcroft, and James M. Harding, Jr. *World Ocean Census: A Global Survey of Marine Life*. Buffalo, NY: Firefly, 2009.

Devey, Colin William, ed. *Oceans*. New York: Springer, forthcoming.

Dinwiddie, Robert, et al. *Ocean: The World's Last Wilderness Revealed*. London, UK: DK, 2006.

Dobbs, David. *Reef Madness: Charles Darwin, Alexander Agassiz, and the Meaning of Coral*. New York: Pantheon, 2005.

Earle, Sylvia A., and Linda K. Glover. *The National Geographic Atlas of the Ocean*. Washington, DC: National Geographic Society, 2001.

———. *Ocean: An Illustrated Atlas*. Washington, DC: National Geographic Society, 2009.

Ellis, Richard. *The Empty Ocean: Plundering the World's Marine Life*. Washington, DC: Island, 2003.

Erickson, Jon. *Marine Geology: Exploring the New Frontiers of the Ocean*. New York: Facts on File, 2002.

Estes, James A., et al., eds. *Whales, Whaling, and Ocean Ecosystems*. Berkeley: University of California Press, 2007.

Fabricius, Katharina. *Marine Life of the Maldives*. Gig Harbor, WA: Sea Challengers, 2001.

Finch, Simon, Kenneth Thomson, and Vincent Gaffney, eds. *Mapping Doggerland: The Mesolithic Landscapes of the Southern North Sea*. Oxford, UK: Archaeopress, 2007.

Fox, William. *Terra Antarctica*. Berkeley, CA: Shoemaker and Hoard, 2007.

Garrison, Tom. *Oceanography*. Belmont, CA: Brooks/Cole, 2007.

Gruber, Nicolas, and Charles Keeling. *Seasonal Carbon Cycling in the Sargasso Sea Near Bermuda*. Berkeley, CA: University of California Press, 2001.

Hekinian, Roger, et al., eds. *Ocean Hotspots: Intraplate Submarine Magnetism and Tectonism*. New York: Springer, 2004.

Hutchinson, Stephen. *Oceans: A Visual Guide*. Buffalo, NY: Firefly, 2005.

Ilyina, Tatjana P. *The Fate of Persistent Organic Pollutants in the North Sea*. New York: Springer, 2007.

Intergovernmental Panel on Climate Change. *IPCC Fourth Assessment Report: Climate Change 2007 (AR4)*. Geneva, Switzerland, 2007.

International Council for the Exploration of the Sea. *Report on the State of the Stock: Cod in the North Sea*. Copenhagen, Denmark, 2006.

Kim, Karl. *Estimating the Impacts of Banning Commercial Bottomfish Fishing in the Northwestern Hawaiian Islands*. Honolulu: University of Hawaii, 2006.

Kirby, Alex. "North Sea Cod 'Face Commercial End.'" *BBC News*, December 16, 2002.

Knox, George A. *Biology of the Southern Ocean*. Boca Raton, FL: CRC, 2006.

Kolbert, Elizabeth. "The Darkening Sea." *The New Yorker*, November 20, 2006.

Langewiesche, William. *The Outlaw Sea*. New York: North Point, 2004.

Leier, Manfred. *World Atlas of the Oceans*. Buffalo, NY: Firefly, 2001.

Lieberman, Bruce. "Changing Ocean Chemistry Threatens to Harm Marine Life." *San Diego Union-Tribune*, September 14, 2006.

Longhurst, Alan R. *Ecological Geography of the Sea*. Burlington, MA: Academic Press, 2007.

Maury, Matthew Fontaine. *Physical Geography of the Sea*. Boston: Adamant Media, 2001.

McGonigal, David. *Antarctica: Secrets of the Southern Continent*. Buffalo, NY: Firefly, 2008.

McLeod, Karen, and Heather Leslie, eds. *Ecosystem-Based Management for the Oceans*. Washington, DC: Island, 2009.

————. *Hawaii Hotspot*. Beau Bassin, Mauritius: Alphascript, 2009.

Miller, Frederic, Agnes Vandome, and John McBrewster, eds. *Maldives*. Beau Bassin, Mauritius: Alphascript, 2009.

Monteath, Colin. *Antarctica: Beyond the Southern Ocean.* Toronto, Canada: Warwick, 2005.

Morato, Telmo, and Daniel Pauly, eds. *Seamounts: Biodiversity and Fisheries.* Vancouver, Canada: University of British Columbia, Fisheries Centre, 2004.

Mundy, Phillip, ed. *The Gulf of Alaska: Biology and Oceanography.* Fairbanks: Alaska Sea Grant College Program/University of Alaska at Fairbanks, 2005.

Myers, Joan. *Wondrous Cold: An Antarctic Journey.* New York: Smithsonian, 2006.

National Oceanographic and Atmospheric Administration. *Magnuson-Stevens Fishery Conservation and Management Reauthorization Act of 2006: An Overview.* Washington, DC: U.S. Department of Commerce, 2007.

National Research Council. *A Century of Ecosystem Science: Planning Long-Term Research in the Gulf of Alaska.* Washington, DC: National Academy Press, 2002.

Nordic Working Group on Fisheries Research. *The Mid-Atlantic Ridge Is Teeming With Life.* Copenhagen, Denmark: Nordic Council of Ministers, March 2005.

Norse, Elliott, and Larry Crowder. *Marine Conservation Biology: The Science of Maintaining the Sea's Biodiversity.* Washington, DC: Island, 2005.

North Pacific Fishery Management Council. *Central Gulf of Alaska Rockfish Demonstration Program.* Washington, DC: U.S. National Marine Fisheries Service, 2007.

North Sea Conference. *The Environmental Impact of Shipping and Fisheries.* Gotenborg, Sweden: North Sea Conference, 2006.

Nouvian, Claire. *The Deep: The Extraordinary Creatures of the Abyss.* Chicago: University of Chicago Press, 2007.

O'Clair, Rita M., Robert H. Armstrong, and Richard Carstensen, eds. *The Nature of Southeast Alaska: A Guide to Plants, Animals and Habitats.* Portland, OR: Alaska Northwest, 2003.

Organization for the Protection of the Marine Environment of the Northeast Atlantic. *Status Report for the Greater North Sea.* London, UK: OSPAR, 2000.

Patton, Kimberley. *The Sea Can Wash Away All Evils: Modern Marine Pollution and the Ancient Cathartic Ocean.* New York: Columbia University Press, 2006.

Pauly, Daniel, and Jay Maclean. *In a Perfect Ocean: The State of Fisheries and Ecosystems in the North Atlantic Ocean.* Washington, DC: Island, 2003.

Pearson, Michael. *The Indian Ocean.* New York: Routledge, 2007.

Pernetta, John. *Guide to the Oceans.* Buffalo, NY: Firefly, 2004.

Pitcher, Tony J., et al., eds. *Seamounts: Ecology, Fisheries, and Conservation.* Malden, MA: Wiley-Blackwell, 2007.

Pollack, Andrew. "A New Kind of Genomics, With an Eye on Ecosystems." *The New York Times,* October 21, 2003.

Rauzon, Mark J. *Isles of Refuge: Wildlife and History of the Northwestern Hawaiian Islands.* Honolulu: University of Hawaii Press, 2001.

Revkin, Andrew C. "Arctic Melt Unnerves the Experts." *The New York Times,* October 2, 2007.

Roberts, Callum. *The Unnatural History of the Sea.* Washington, DC: Island, 2007.

Robinson, Ian. *Measuring the Oceans From Space.* New York: Springer, 2004.

Rogers, Alex D. *The Biology, Ecology, and Vulnerability of Seamount Communities.* Gland, Switzerland: World Conservation Union, 2004.

Rose, Paul, and Anne Laking. *Oceans.* Berkeley: University of California Press, 2009.

Rozwadowski, Helen M. *Fathoming the Ocean: The Discovery and Exploration of the Deep Sea.* Cambridge, MA: Belknap Press, 2005.

Scott, Jonathan, and Angela Scott. *Antarctica: Exploring a Fragile Eden.* New York: HarperCollins, 2008.

Sharp, Renee, and Rashid Sumaila. "Quantification of U.S. Marine Fisheries Subsidies." *North American Journal of Fisheries Management* 29 (2009): 18–32.

Shirihai, Hadoram. *The Complete Guide to Antarctic Wildlife.* Princeton, NJ: Princeton University Press, 2008.

Siedler, Gerold, et al., eds. *Ocean Circulation and Climate: Observing and Modelling the Global Ocean.* San Diego, CA: Academic Press, 2001.

Smith, Roff. *Life on the Ice.* Washington, DC: National Geographic Society, 2005.

Smith, Struan. *Bermuda: Environment and Development in Coastal Regions and in Small Islands.* St. George, Bermuda: Bermuda Biological Station for Research, 2000.

Soper, Tony. *The Arctic Ocean.* Guildford, CT: Globe Pequot, 2001.

Spies, Robert. *Long-Term Ecological Change in the Northern Gulf of Alaska.* Amsterdam, The Netherlands: Elsevier Science, 2007.

Steinberg, Philip E. *The Social Construction of the Ocean.* New York: Cambridge University Press, 2001.

Sverdrup, Keith, and E. Virginia Armbrust. *An Introduction to the World's Oceans.* New York: McGraw-Hill Higher Education, 2008.

Takahashi, Eiichi, et al., eds. *Hawaiian Volcanoes: Deep Underwater Perspectives.* Washington, DC: American Geophysical Union, 2002.

Tove, Michael H. *Guide to the Offshore Wildlife of the Northern Atlantic.* Austin: University of Texas Press, 2001.

Trujillo, Alan P. *Essentials of Oceanography.* Upper Saddle River, NJ: Prentice Hall, 2007.

United Nations Food and Agriculture Organization. *State of World Fisheries and Aquaculture 2008.* Rome, Italy, 2008.

Weir, Gary E. *An Ocean in Common.* College Station: Texas A&M University Press, 2001.

World Conservation Union (International Union for Conservation of Nature). *The IUCN Red List of Endangered Species.* Gland, Switzerland, World Conservation Union (IUCN), 2009.

Worm, Borris, et al. "Impacts of Biodiversity Loss on Ocean Ecosystem Services." *Science* 314 (November 2006): 787–790.

Index

Italic page numbers indicate images and figures.

Milton Keynes UK
Ingram Content Group UK Ltd.
UKHW052027141024
449569UK00016B/731